INTERFERENCE HANDBOOK

William R. Nelson, WA6FQG
Editor: William I. Orr, W6SAI

Dedicated to Rusty and Jan and the others who gave encouragement and assistance

Caution: Working on antennas or towers can be dangerous. All warnings on the equipment and all operating and use instructions should be adhered to. Make sure that the antenna is disconnected from the station equipment before you begin to work on it. Make sure that your antenna is not close to power lines, and that it cannot drop on a power line if wires or supports fail. Do not attempt to climb a tower without a safety belt. It is best to work on your antenna with someone who can assist you with tools and will be able to help in the event of a problem or an emergency.

Published in 1993 by Radio Amateur Callbook
(an imprint of Watson-Guptill Publications,
a division of BPI Communications, Inc.),
P.O. Box 2013, Lakewood, New Jersey 08701

Library of Congress Catalog Card Number: 81-51709
ISBN 0-8230-8709-3

Manufactured in the United States of America

1 2 3 4 5 6 7 8 9/01 00 99 98 97 96 95 94 93

TABLE OF CONTENTS

FOREWORD

Welcome to the Wonderful World of Interference!

Bill Nelson, the author of this Handbook, has been involved with the interference problems of radio amateurs, CB operators and users of home entertainment devices for the past sixteen years as the Amateur Radio Representative of the Southern California Edison Company. He's been a detective, investigator and mediator for thousands of interference problems encountered in the southern California area. His experience in interference of all types, and his uncanny ability to trace sources of interference, make him the undisputed expert in this field.

The knowledge gained in his work is incorporated in this Handbook. The stories he tells are true as he has either been involved in them or has learned about them from other RFI investigators.

The purpose of this Handbook is to help others locate and resolve interference problems of every type. Sources of interference are described along with the methods used to locate them. Suppression circuits for interfering devices are discussed in detail as well as protection techniques for home entertainment equipment.

Interference is a fast-growing problem and there will be more tomorrow than there is today. The author and publishers hope that this Handbook will be of service to those confronting this worldwide obstacle to clear, reliable radio communication.

The Author wishes to thank the following for help in the compilation of this Handbook:

Wilbur Bachman, W6BIP
Stuart Cowan, W2LX
Drew Diamond, VK3XU
William Orr, W6SAI
Phillip Rand, W1DBM
William Steward, K6HV
Robert Coe, Senior Vice President (retired) Southern California Edison Co.
Marvin Loftness, Capitol Consultants, Olympia, WA
Santa Barbara (CA) Amateur Radio Club (TVI Committee)
Electronic Industries Association (Consumer Electronics Group)
Federal Communications Commission
Institute of Electrical and Electronic Engineers, Inc.
Wireless Institute of Australia
Florida Light and Power Company
Champion Spark Plug Co.
HyGain/Telex Communications Corporation.
Southern California Edison Co. (RFI Investigators and Communication Technicians)
Southern California Gas Co.

For the ARRL RFI Assistance List, special thanks to:

Dick Baldwin, W1RU
Laird Campbell, W1CUT
Dave Sumner, K1ZZ
Harold Richman, W4CIZ

Chapter 1

An Introduction to RFI
--Radio Frequency Interference--
It's All Around Us!

What About RFI?

*Radio Frequency Interference (RFI)*was born the day the first radio transmissions were made by Marconi. One of his greatest problems was the reduction of interference between two broadly tuned spark transmitters (Figure 1). If Marconi were alive today he would understand the complexity of RFI in today's world of instant electronic communication.

In the early days of radio it was called broadcast interference (BCI). As television boomed in the "fifties", it was called television interference (TVI). And today we have radio frequency interference (RFI) that affects all sorts of electrical and electronic devices that were largely unheard of a decade ago. Early BCI was manageable because it affected only a few listeners; the TVI problem was much greater because of the explosion in tv viewing when inexpensive black-and-white sets became available. RFI, on the other hand, is a national problem that affects all individuals to a greater or lesser degree. The sources of RFI have proliferated and devices that are susceptible to RFI abound. Look at these numbers. In the United States in 1980 there were over--

8,200 radio stations (broadcast and fm)
970 television stations
15,000,000 CB transmitters
360,000 amateur radio stations
210,000 aviation transmitters
7,800 radar transmitters
300,000 industrial radio transmitters
115,000 police and fire department radio transmitters
36,000,000 two-way portable radio transmitters

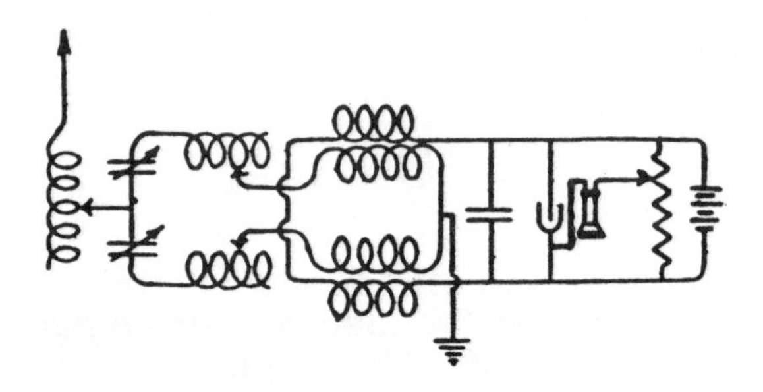

Fig. 1 An early attempt at reducing interference. This 1917 circuit was supposed to send signal and interference on separate paths in the crystal set. The desired wavelength was detected but the unwanted wavelength was rejected. Circuit failed to perform as interference and signal were usually on the same frequency. (From "Manual of Radio Telegraphy and Telephony", U.S. Naval Institute, 1918).

Plus millions of microwave ovens, X-ray machines, electric motors, light flashers and dimmers, welding machines, neon signs, diathermy machines, and so on.

All of these are potential sources of radio interference.

And that's not all. Radio and television receivers themselves can cause RFI, and in the United States in 1980 there were over 413,000,000 radio receivers and 125,000,000 television sets. And while you read this countless thousands more radio receivers and transmitters are being bought and used.

Well, you might ask, what has all of this got to do with radio or television interference? What difference does it make whether all of these radios are properly used? What's the problem?

The problem simply is this: *all* radio receivers and *all* transmitters are potential sources and victims of RFI for a number of reasons. The receiver or transmitter may be poorly designed, improperly built, incorrectly installed or badly tuned. Any or all of these factors can create unwanted interference in a nearby radio, television or stereo set; to amateur radio and CB operators; and to navigational, public service and government radio services.

In addition, almost *anything* run by electricity can create RFI. Such everyday things as light bulbs, automobiles, electric blankets, electronic games, mini-computers, coffee pots, washing machines, intrusion detectors, electric razors, door bells--you name it--are potential sources of

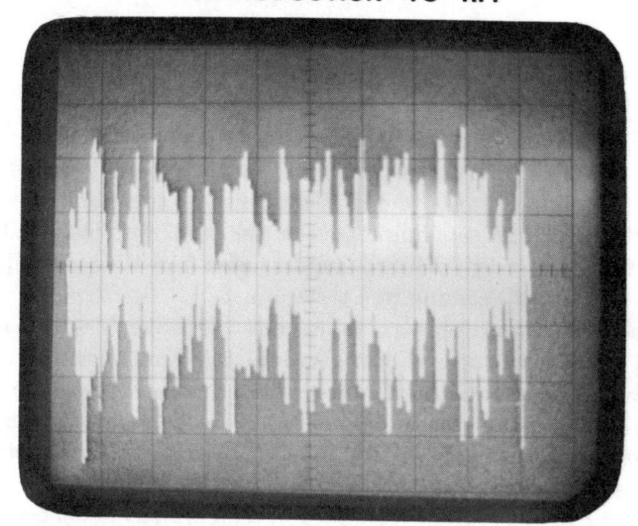

Fig. 2 Typical radio frequency interference.

interference. Whenever a gasoline motor turns over a spark is created to fire the plug and a radio noise is generated. Many ignition systems radiate a blast of radio noise over a broad chunk of the radio spectrum (Figure 2).

Flickering neon signs add their bit to the racket and even the food you cook in a microwave oven may contribute its share of RFI in the microwave portion of the spectrum. Almost any machine, even a mechanical one, can create RFI.

And to top it all off, it is estimated that there are nearly 500,000 additional illegal radio transmitters in the United States (most of them operating in the CB channels).

No wonder that electronic bedlam surrounds us.

And worst of all, heavy equipment and power lines of electric utilities can create severe RFI which can be propagated for miles along the lines. It's a wonder that *anybody* can hear *anything* on the radio or see *anything* on television considering the vast number of interference-generating devices that surround us!

By fall, 1979, the number of annual interference complaints received by the Federal Communications Commission exceeded 70,000. And the Commission is bracing itself for an expected 100,000 complaints by 1980. Most of the complaints involve impaired television reception, followed by interference to radio receivers and stereo equipment. And, unfortunately, a large percentage of all interference complaints are associated with CB radio transmissions.

History of RFI

Electromagnetic interference first became a problem in the early days of the telegraph and telephone. By 1885 the network of telegraph and telephone lines routed together on common poles created interference and coupling between the two systems. And by 1890 interference on telephone circuits was noted from the new dc electric railways ("trolleys") coming into general use.

By 1900 it was necessary to separate telephone, telegraph and power lines because of interfering noise and the first investigations of inductive power line interference and studies of good construction practices were underway. The Radio Act of 1912 recognized the problem of radio interference for the first time but offered no solution other than to suggest that the radio transmitters should emit "pure waves".

After World War I telephone and telegraph systems became much more complicated; dialing pulses and other special circuits were added and to resolve interference between circuits the International Telephone Consulting Committee was created under the auspices of the League of Nations.

Radio communication developed rapidly in the "twenties" and quickly progressed into a number of rather sophisticated systems without much thought being given to problems of interference. Because of the interference problems caused by frequency congestion and poor control of signals and harmonics, the federal government established the Federal Communications Commission (FCC) in 1934 to regulate the use of radio and wire communications. Though the authority of the FCC is primarily limited to the user of communication devices, it is continually seeking to eliminate some of the interference developed by electronic devices.

RFI and the FCC

The Federal Communications Commission has been controlling the generation of electrical interference to some extent for many years. As communications circuits and equipment become more complex and utilize a larger portion of the spectrum, and the spectrum becomes more crowded with radiation from electronic, industrial, commercial and consumer equipment, the FCC attempts to increase the scope and effectiveness of its regulations. FCC rules covering RFI implement a 1968 law empowering the FCC to make reasonable regulations governing the

Type of consumer device	Part	Type of consumer device	Part
Wireless intercom	15.7	Microwave oven	18
Walkie-Talkie (under 100 mW)	15.119	Electronic games	15.7
Wireless microphone	15.161	Class 1 TV device (couples to TV set with modulated rf oscillator)	15-H
Security transmitter	15-D	Master antenna system	15.13
Intrusion detector	15-F	Electronic equipment	15.13
Field disturbance sensor	15-F	Radio receivers (above 30 MHz)	15-C
Door opener transmitter	15-E	Electric motors	15.25
CB transmitter	95	Home appliances	15.25
Model remote control transmitter	95	Ignition system	15.25
Amateur radio transmitter	97	Light dimmer	15.25

Table 1. A portion of the FCC Rules and Regulations dealing with RFI.

interference potential of certain devices. The purpose of the rules is to require compliance with equipment standards by manufacturers, importers and distributors of rf devices, as well as by users.

Rf devices subject to FCC authority range from radio transmitters of all types to restricted radiation devices such as receivers, wireless microphones, radio garage door openers, tv games and toys. Also included are industrial, scientific and medical equipment. Exempt from the laws are devices manufactured solely for export and devices to be used by the U.S. Government. Marketing rules and type approval have been defined for certain equipments capable of causing RFI. These controls are summarized in Parts 15 and 18 of the FCC Rules and Regulations, available on a subscription basis from the Superintendent of Documents, U.S. Government Printing Office, Washington, D.C. 20402. This data is included in Volume II and is not sold separately.

Consumer Devices and the FCC Rules

Since the time the FCC regulations governing consumer electronics were written between 1946 and 1948 vast technological changes have occurred. Semiconductors, integrated circuits and digital systems have appeared and many new devices, operating at frequencies substantially higher than in the late 40s, present new interference problems. The FCC has, on occasion, amended or added to the original regulations and new changes are in the offing. Notices of Inquiry have been posted to study new rules dealing with RFI. A summary of the Regulations covering various devices is listed in Table I.

Many broadband interference sources such as electric motors, home appliances, light dimmers and ignition systems are disruptive to electronic communications systems. These devices, with the exception of auto ignition systems, are quasi-covered by Part 15 (Incidental Radiating Devices) of the FCC rules.

A number of FCC studies, however, have been initiated to examine the advisability of establishing FCC rules for interference limits. For example, the FCC measured the ignition radiation levels of over 10,000 vehicles at land-mobile communication frequencies. The Commission decided that objectionable degradation of communications had occurred, particularly in high density urban traffic. The Motor Vehicle Manufacturer's Association of the United States responded to this inquiry by questioning the validity of the FCC's test program of the 10,000 vehicles.

Many manufacturers oppose mandatory FCC standards but little hope for FCC imposed RFI standards exists until the Communications Act is amended or otherwise modified by Congress.

RFI-What To Do About It? Two Important Facts

Yes, RFI is Hell if you have it. The purpose of this Handbook is to outline the many sources of interference; explain how to eliminate or reduce them; and tell you how to protect yourself against RFI. The causes and cures of RFI are discussed in nontechnical language that is easy to read and understand. This Handbook also helps you to operate your ham or CB station free of annoying RFI to your neighbors and helps you to enjoy good listening and happy viewing, whether you are ham, CBer or an owner of electronic equipment tormented by interference.

Remember--all cases of RFI involve two things: the *source* of the interference and the *victim* of the interference. For a complete RFI cure, the interference must be suppressed at the source, and the victim (the receiver or stereo equipment, or whatever) should be protected, or modified, in such a way as to reject interference.

Meanwhile, the tide of interference pollution mounts with each passing day. The continuing upsurge in the number and kinds of electronic equipment grows. Every new device has the potential to create RFI, or of being interfered with.

Interference--The Harm It Can Do

Interference rears its ugly head in many forms. It can be the squiggly lines on your television screen, the strange voice on top of Beethoven's Ninth on your stereo, or a ghostlike noise on the fm signal. It can be the vacuum cleaner talking back to the housewife, the voice coming out of the electronic organ or other equally baffling manifestations. Generally speaking, aside from thunderstorms and lightning static, most interference is man-made and can be located and suppressed. The problem is finding the source and then applying the remedy. That's not so easy.

RFI may originate within your own home, within your neighbor's house, from a mobile radio in a fast-moving vehicle passing your house, from nearby utility lines, or from unknown sources a thousand miles away. It can invade your electronic equipment through the antenna circuit, or sneak stealthily in via the power cord. In severe cases, it can be picked up directly by the wiring of the house, or the cables and wiring of your equipment.

Aside from the irritation of having your reception ruined, your TV picture disrupted or your stereo blasted, is there any truly dangerous effect of RFI? Yes, there is!

Consider an RFI-induced malfunction in an electronic heartpacer implanted in a heart patient. Or, think about the failure of the electronic braking system in an automobile at a crucial moment in traffic. Imagine the results of the failure of the electronic guidance system in a jumbo jet packed with passengers landing in a heavy storm. All of these dangerous situations have come to pass because of equipment malfunction caused by external RFI.

Luckily, most people haven't been exposed to such dangerous situations, but they do exist and more and more RFI problems are found every day as our world becomes more complex and as we invent electronic "black boxes" to perform new tasks for us.

Is Anything Being Done About RFI?

The inability of most electronic equipment to operate without disturbing another piece of electronic equipment has been known for decades. Numerous programs to study *electromagnetic compatibility* have been undertaken by the military and by private industry. Much has been done to eliminate or reduce the effects of RFI in expensive military and aeronautical equipment; unfortunately, little of this thinking has trickled down into the less-expensive, highly competitive consumer market.

Manufacturers of home entertainment equipment (TV, radio, stereo, record players, electronic games, etc.) believe that only a small percentage of the total sets sold will ever be used under conditions of RFI and are reluctant to design in interference rejection circuits that may never serve a purpose. There is little incentive for one manufacturer to incorporate RFI protection circuits and shielding in his product if it places him at a price disadvantage with his competition.

The manufacturers of transmitting and industrial power equipment, in a similar manner, only include enough RFI filtering and suppression in their gear to satisfy the minimum requirements of the Federal Communications Commission. Obviously, a wide gap exists between these two design philosophies.

RFI Legislation

From time to time, bills giving the FCC the authority to set standards covering RFI susceptibility of home entertainment electronic devices have been introduced before Congress. To date, none of these bills has been passed for various reasons. While the FCC has determined that most RFI problems can be lessened or cured at the point of reception, the FCC has no power at present to regulate passive electronic devices that can experience RFI.

Some manufacturers supply filters for reducing television and stereo interference when the problem is brought to their attention. But on the whole, it would seem that the only means of insuring that the nontechnical public can have home entertainment equipment capable of operating properly in the vicinity of nearby sources of radio energy is to require the inclusion of interference protection circuits in the consumer electronic devices. To date, this legislation has not come to pass.

The Wide, Wide World of RFI

In addition to radio transmitters and sparking devices, severe problems of RFI can be created by power lines, heavy electrical machinery, power substations and large radio frequency generators used in the manufacture of plywood, paper and plastic products.

Worse still, some objects such as power pole hardware, antenna guy wires, inoperative receivers attached to an antenna, and even more unimaginable objects can create RFI where none existed before. The whole problem is complex and confusing, especially to the proud owner of a new tv or stereo receiver who experiences interference. He is quick to fix the blame on the CBer or radio ham down the block as he associates the use of radio transmitting equipment and antennas with interference. Alas, the radio operator takes the blame while in many instances he is blameless, and possibly suffering from the same interference. But this is difficult to explain to an unhappy viewer or listener.

What does RFI mean? Radio frequency interference is a broad term that covers any type of electrical signal capable of being propagated into, and interfering with, the proper operation of other electrical or electronic equipment. Another term for RFI is *electromagnetic interference* (EMI); this latter term is much favored by the military. But both terms, in essence, mean the same thing.

As already mentioned, RFI can originate in almost anything that uses electricity or that is connected to a power line. The following is a partial list showing some potential sources of RFI:

In industrial and business areas:

Air conditioners
Annunciator systems
Arc welders
Belt static
Broadcast stations
Cash registers
Circuit breakers
Culture incubators
Dental drills
Dental sterilizers
Diathermy machines
Distribution systems
Electric commutating motors
Electric fences
Electric railways
Elevators
Flashing signs
Fluorescent lamps and starters
French fryers
Gasoline engines
Germicidal lamps
Ignition systems
Induction heaters
Intrusion detectors
Lightning arresters
Linear accelerators
Mercury arc rectifiers
Neon signs

Office machines
Oil furnaces
Pest control devices
Power lines and poles
Plastic molding equipment
Printing presses
Relays
Rf heaters and welders
Radio telescopes
Radio transmitters (all types)
Refrigeration equipment
Security transmitters
Shoe repair machines and belts
Sign flasher buttons
Soldering machines
Spot welders
Sterilizers
Synchronous converters
Television and fm stations
Thermostats
Tractors
Trucks and vans
Ultraviolet systems
Ventilating fans
Waffle cookers
Walkie-talkies
X-ray machines

And hundreds of other devices you can think of!

In addition to many of the above, the following RFI sources are found in residential areas:

Air conditioners
Air purifiers
Amateur and CB transmitters
Automobiles and trucks
Bottle warmers
Butter keepers
Christmas decorations
Community tv antenna systems
Defective sockets and lamps
Deodorizing lamps
Door opener transmitters
Doorbell transformers
Electric blankets, fences, games
Irons, razors and tools
Electronic games
Fish tank thermostats
Flashers
Food mixers
Furnace controls
Garbage disposals
Gas furnace blowers
Gas range and dryer ignitors
Hair dryers
Heating pads
Home elevators
Light dimmers
Loose fuses and lamps
Movie projectors
Power lines
Radio and tv receivers
Scanner receivers
Sewing machines
Slot car racers
Stereo systems
Traffic control switches
Vacuum cleaners
Washing machines
And more and more!

And throughout communities run power lines which can be prolific sources and carriers of radio noise, sometimes combining various noise sources into one gigantic roar which makes the RFI solution more complex than it might otherwise be.

It can be seen that while a nearby radio ham, CBer or other radio station may be causing RFI, there are numerous other noise sources in the vicinity that can be annoying generators of interference. By listening to, or looking at, the RFI and judging its characteristics, it is often possible to determine the source, if not the exact location, of the annoying interference. In the great majority of cases RFI is unintentional, that is, the operator of the offending equipment doesn't know he is the source of the interference.

In addition to unintentional noise interference, some countries indulge in intentional interference, or "jamming", of the radio programs that are politically unacceptable to them. The Soviet-bloc countries, in particular, are often heavy users of jamming stations which blot out large chunks of the various broadcast bands in an attempt to prevent their citizens from listening to broadcasts which may give them something to think about. This odious form of interference comes and goes depending upon international tensions.

Natural Sources of Interference

Added to the unholy din of man-made RFI must be the natural sources of radio noise: static, lightning, atmospheric hiss, radio noise from outer space, and the noise generated by the movement of electrons in the heart of all electronic equipment. In short, our world is filled with RFI both natural and man-made, and as our culture advances, the level will increase--unless something is done about it now.

Noise limits the useful operating range of all radio equipment. Aside from man-made noise, natural radio noise is all around us. A major source of *atmospheric noise* (static) is a belt of continuous lightning activity around the equator from which interference is propagated toward the rest of the world by ionospheric reflection (Fig. 3). Thunderstorms in the temperate zones create additional radio noise; the overall noise level depends upon frequency, time of day, season of the year and geographical location.

A second source of natural interference is *galactic noise* from outer space. A major generator of galactic noise is the sun and various star groups have been identified as radio noise sources. Additional galactic noise seems to come from areas of space which are thinly populated by stars. The source of this radio noise is unknown.

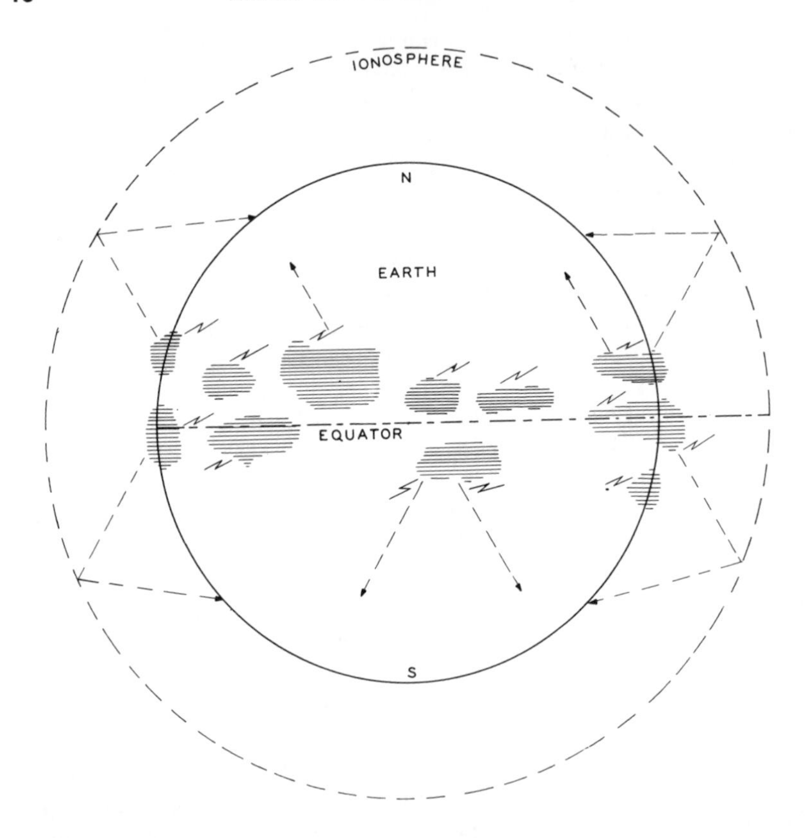

Fig. 3 Major source of atmospheric noise is a belt of continuous lightning activity around the equator from which interference is propagated toward the rest of the world by ionospheric reflection.

Atmospheric noise predominates on the lower radio bands and is relatively unimportant above 20 MHz. *Sky noise,* on the other hand, which is created by agitation of the molecules of the atmosphere by sunlight, is low below 500 MHz and increases with frequency. Galactic noise extends from about 15 MHz upwards in frequency, being limited at the low frequency end of the radio spectrum by atmospheric absorption.

Precipitation static is often found in the hf spectrum during a heavy rainstorm, showing up as a loud hiss in the radio receiver or as "snow" on the screen of a tv set. It is caused by static electricity building up on the receiving antenna.

Finally, there is a low level of *thermal noise* in all electrical equipment caused by the thermal agitation of electrons in conductors.

All of these noise sources, plus man-made noise, form a "noise floor" in the communication spectrum below which the sensitivity of the communication equipment is restricted. And in the long run, a receiver's ultimate sensitivity, in the absence of man-made noise, is determined by natural noise. In the rare instances where external natural noise is low or absent, the ultimate noise level is determined by the noise level inherent in the equipment generated by the movement of electrons as they go about their mysterious work in the conductive circuits.

Noise Measurement

It is possible to measure noise. The noise level of radio equipment can be expressed in various ways and one of the most convenient expressions of noise is to reference the noise to the thermal agitation of electrons. The noise is measured in a radio receiver over a specified segment of the radio spectrum and the noise power is expressed in terms of bandwidth referenced to a noiseless condition.

Thermal noise itself is expressed as follows:

$$E = (4\ kTR \cdot \Delta f)^{1/2}$$

where, E is the rms noise voltage

R is the resistive component across which the noise is measured

k is Boltzmann's constant (1.38×10^{-23} joules/degree Kelvin)

f is the bandwidth in Hertz

T is the temperature in degrees Kelvin

Thermal noise, thus, is a function of temperature and bandwidth and some very sensitive receiving devices have taken advantage of this by restricting bandwidth and operating immersed in liquid nitrogen at near-absolute zero temperature.

External noise can be measured with much simpler equipment, such as an "all-band" communications receiver and the source of noise can often be pinpointed by the use of either a directional receiving antenna or a mobile receiver installation. Simple equipment that will do the job is discussed in detail in this Handbook.

The Practical Aspects of RFI

The natural "radio noise floor" is only of academic interest to most radio enthusiasts as the general level of urban RFI is many times higher than the natural noise level, at least below 50 MHz. Worldwide, the radio spectrum is increasingly crowded with communication circuits and the radio noise pollution level continues to rise as more and more sources of RFI are placed in service with little or no control over their potential for interference to other radio circuits.

RFI control has been an unrewarding, uphill battle. Tighter control over manufacturing standards and rigorous inspection of communication equipment is helping to contain RFI pollution to a degree, and future legislation can assist in minimizing or eliminating many types of interference. Gains in this field, however, are partially offset by the explosive growth in consumer electronics and the increasing need for worldwide communication and data exchange. Deliberate interference (jamming), too, is a vexing problem that is not openly acknowledged.

What Equipments are Susceptible to RFI?

An accounting of all the devices that are susceptible to RFI is impossible as the list grows daily. The following partial list will give you an idea of the all-encompassing effect of RFI upon our daily lives:

Broadcast, fm and tv receivers
CATV systems
Closed circuit video systems
Computers and terminal equipment
Digital timers
Electro-explosive devices
Electronic ignition systems
Electronic measuring equipment
Electronic musical equipment
Electronic sensors
Home audio and stereo equipment
Metal detectors
Medical monitoring equipment
Process control devices
Radio controlled equipment
Telephone switching equipment
Vehicle ignition and braking systems

EMP-The Ultimate RFI Problem

Electromagnetic Pulse Generation (EMP) is a recent RFI problem and may be the ultimate and final pollution source. EMP is the electro-

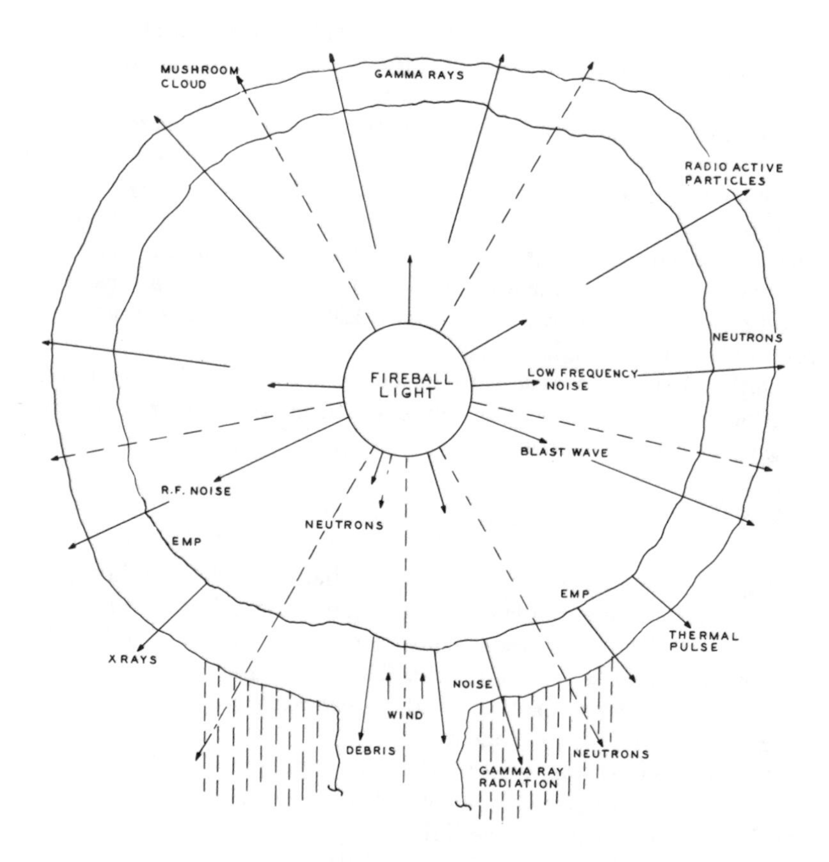

Fig. 4 Nuclear explosion is a prolific source of electromagnetic energy. The primary source of RFI is gamma radiation. Such an explosion in the atmosphere also destroys ability of the ionosphere to reflect radio waves.

magnetic "shock wave" resulting from a nuclear explosion. Many types of modern communication equipment, particularly solid state gear, are susceptible to high voltage transients and nuclear radiation damage, both of which are a direct result of a nuclear explosion. It is not inconceivable that such an explosion, high in the atmosphere, could create long range EMP to knock out an adversay's communication and guidance system electronics.

In addition, tests have shown that a nuclear explosion high in the atmosphere destroys the ability of the ionosphere to reflect and refract high frequency radio waves, thus these circuits are rendered useless for a period of time by an A-bomb or H-bomb explosion. The first such tests were run in the late 1950's and the effects noted by radio amateurs and the military.

Within the first second after detonation, a nuclear blast releases all of its energy, producing invisible, penetrating and harmful rays and particles in the form of alpha and beta particles, gamma rays, x-rays, neutrons, electrons and neutrinos (Fig. 4). The primary source of electromagnetic pulse generation is gamma radiation, traveling outwards from the blast in all directions. The EMP from an atmospheric burst covers large geographical areas, a 1200 miles radius being typical for a detonation at high altitude.

Military studies are continually being undertaken to devise methods of protecting communication equipment from the effects of electromagnetic pulse generation.

The "Woodpecker"

High frequency radio communication has been plagued during the past years with an annoying form of man-made interference termed *the woodpecker.* The interference consists of a series of rapid pulses that occupy large chunks of the spectrum. The pulses are extremely powerful and seemingly move about at random, knocking out communications over a wide swath of frequencies.

Examination of the pulses has led to the conclusion that they are generated by a variable frequency backscatter radar system located in the Soviet Union. The purpose of the radar, it is assumed, is to detect the launch of intercontinental ballistic missiles.

The system seems to have two transmitters located in the Ukraine and Siberia that transmit pulses over the North Pole towards the United States. The reflected pulses provide information on ionized trails left by the launch of a surface missile.

Complaints have been in vain--the "Woodpecker" remains on the air--at least up to the time of the publication of this Handbook. It is rumored that additional hf radars will be in operation during the coming decade, adding more noise and interference to an already overburdened radio spectrum.

Chapter 2

Spark Discharge Interference

Interference Transmission and Spark Discharge

Electromagnetic interference (RFI) can be transmitted by *radiation, induction* or *conduction.* Radiation is electromagnetic propagation through space. Conduction is transmission through an electrical circuit. The most common example of this type of transmission is the passage of interfering signals through power and control leads of equipment.

Figure 1 illustrates the various means by which interference is transmitted between systems. The source of interference radiates interference directly towards the receiver and also conducts interference back through the power source. If the receiver is unshielded or improperly shielded, it will pick up the radiated interference. Interference from the source is also conducted through the power line and radiated from the line to the receiver and induced by the proximity of source and receiver power lines.

RFI can be cured in the majority of cases by a systematic investigation of the noise source and transmission path. Many radio amateurs and CBers have traced RFI when neighbors have complained of impaired radio or tv reception. A growing business field is the detection and elimination of RFI and some public utility companies have a staff whose responsibilities are tracking down and eliminating RFI for the service area of the utility. Unknown a few years ago, the job of the *RFI Investigator* is a new craft created by the growing problems of radio interference in the large metropolitan areas of the United States.

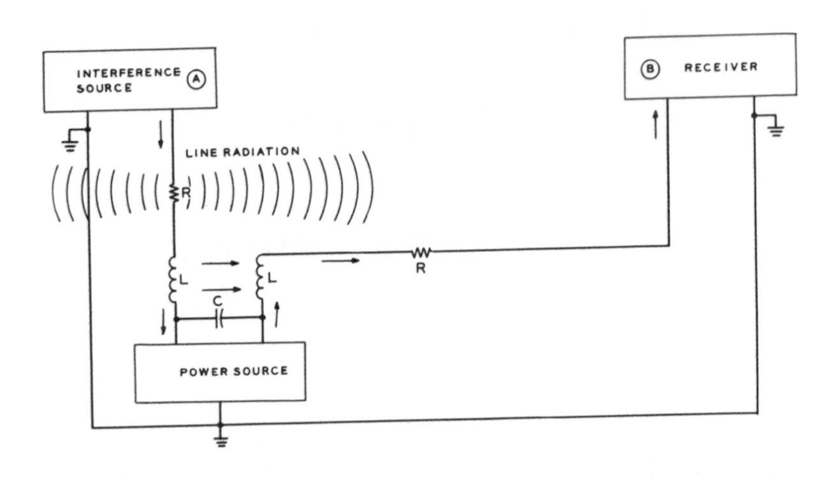

Fig. 1 Transmission paths of RFI. Interference can be radiated directly from the interference source A to the receiver B or it may be coupled via the power lines or through the power source. Coupled inductance and capacitance of lines provides a ready path for interference as does the common impedance of the supply.

The RFI Investigator

The writer of this Handbook has a unique occupation. He is one of a handful of RFI investigators for the general public. He is employed by a large electric public utility in California where the expanding population and rapidly growing industry have created a myriad of RFI problems. He has been intimately involved with thousands of RFI complaints made to the power company and the FCC by unhappy viewers/listeners who are having their reception disrupted by interference. He has learned the hard way that the public reaction to RFI is quick and outspoken but their knowledge of the problem is zero. His job is to investigate the RFI complaint, track the interference to its source and resolve the problem. His radio amateur and CB friends call him "The Investigator" and the name has stuck.

Years of experience have taught the investigator the technique of locating a noise source, or other interference problems. The investigator has a car full of receiving equipment that can be used to track noise sources over the spectrum range of 540 kHz to 450 MHz. Much of the

Fig. 2 Mobile installation used by the author to hunt RFI. The equipment includes a Collins all-wave receiver, a 51S-1 with a 2 and 6 meter converter and a 2 meter fm transceiver. Also included is a portable am-fm broadcast receiver. The equipment is mounted in a removable rack installed next to the driver's position.

equipment is ham gear that can be found in modern ham shacks (Figure 2). No special directional antennas or loops are required.

An important tool in such an investigation is a portable battery operated radio that can be used as a hand-carried direction finder to track the noise source.

Locating the Noise Source

A later chapter discusses the equipment and techniques you can use in locating interference sources. Generally speaking, the approximate direction of the noise source is determined by the use of a portable, multiband, battery operated receiver having a built-in loopstick antenna. The directional properties on the broadcast band of this receiver will indicate a line of position along which the interference

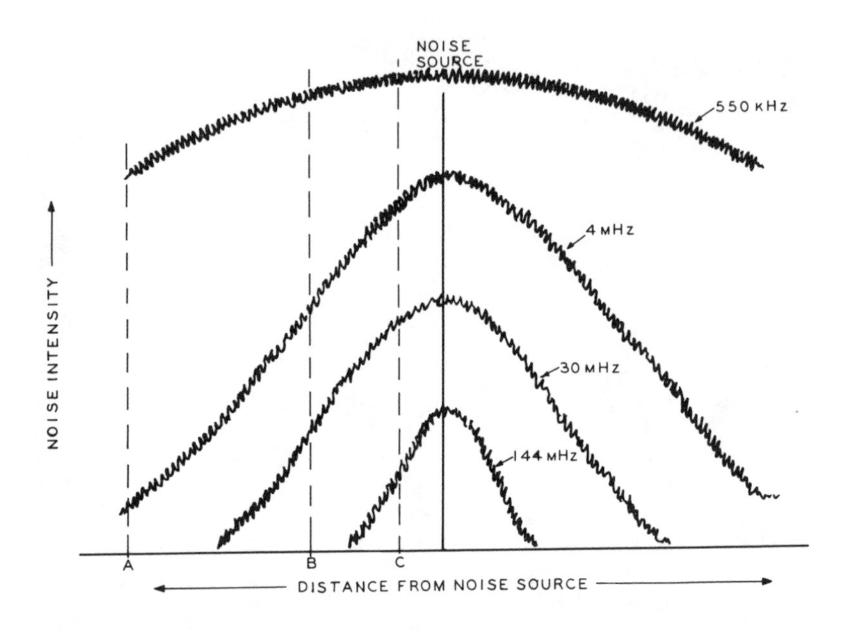

Fig. 3 The closer the noise source, the higher in frequency it may be heard. At location A noise is very loud at 550 kHz but can barely be heard at 30 MHz. At location B (closer to the noise source) the noise is much louder at 30 MHz. At C (very close to the noise source) the noise can be heard at 144 MHz. A portable fm receiver (88-108 MHz) is very handy in locating noise sources.

source probably lies. Driving about an area "infested" with noise usually provides a general indication of the noise source. Experience plays a large role in this initial stage of the investigation.

The investigator knows that the distance covered by interference is inversely proportionaly to the frequency of reception. That is, the closer he gets to the noise source, the higher in frequency it can be heard (Figure 3). Thus, while listening a great distance from the interference source he might only hear it at the "top" of the broadcast band near 550 kHz. As the investigator draws nearer to the source, he might be able to hear it at 4 MHz. Drawing still closer, the noise may be apparent at 14 or 28 MHz. At this point, the search is shifted into the fm broadcast band (88-108 MHz) or perhaps into the 144 MHz amateur band.

Interference caused by amateur and CB transmitters is tracked in much the same manner although tracking is done on the frequency (or harmonic frequency of the transmitter). Call letters, if given, are an important clue to station location. And even if the transmissions are from an unlicensed station, the same technique will work in tracking down the offending source. Perseverance, experience and several receivers in the automobile can work wonders in the interesting and unusual occupation of RFI investigation.

The best way to explain the technique of RFI location and elimination is to start by studying real-life cases of RFI that are commonly encountered, and which have been solved by common sense and the cooperation of the individuals involved. The remainder of this chapter is devoted to a look at some interesting RFI noise sources, their characteristics and the means used to eliminate them.

RFI does not limit itself to radio reception, as the term implies. In addition to tv and stereo interference, RFI includes telephone and hearing aid interference and interference to a thousand other electrical devices. RFI is a catchall term and anything that conveys electrical intelligence to man is subject to RFI. The possibilities are endless.

RFI signifies that an unwanted electrical signal or disturbance is present in a circuit in sufficient strength to be heard or seen, or to otherwise disrupt the presence of the desired signal. It is highly subjective because what is severe RFI to one person is virtually unnoticed by another.

An RFI Noise Source--The Spark Discharge

A partial list of equipment capable of creating RFI, particularly of the spark discharge type is given in Chapter 1. You can think of other devices which fit this category.

In addition to the spark discharge, there are two other classes of rf noise generators: the electrostatic discharge and the rf oscillator. Radio transmitters are included in this last class. Each class of noise has its own peculiarities and leaves its own unique "noise print", a fact that is helpful in tracking down the source. Let's examine these sources, starting with spark discharge, which is common to all neighborhoods.

The spark discharge is a luminous, disruptive electrical breakdown of short duration (Figure 4). A natural discharge of this type is a lightning stroke. A good example of useful spark discharge is the ignition system and spark plugs in a gasoline motor.

The characteristic sound associated with a spark discharge may variously be described as a buzzing, rasping, popping noise similar to bacon frying in a pan. Turn on an old motor-style electric razor and you'll hear this characteristic noise in a nearby receiver.

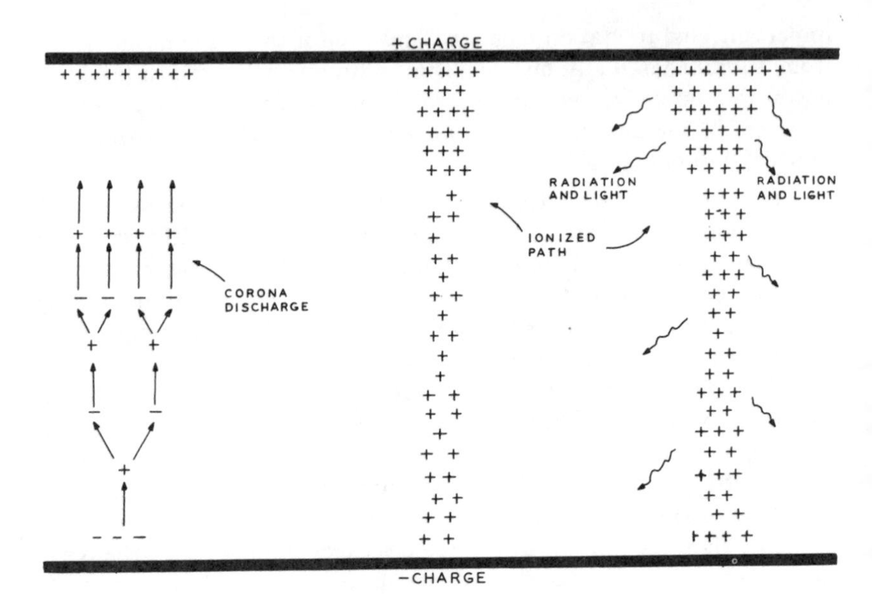

Fig. 4 Start of spark discharge. The spark is a transient form of gaseous conduction caused by ionization of air between charged surfaces. At left, the electric field produces corona discharge and ionization of air molecules. A stream of positively charged ions gains energy from the electric field and builds into an avalanche, as shown at center. At right, the mature ion avalanche passes from the negative to the positive surface. Electrons freed by avalanche action will produce further avalanches through the region so that the entire path becomes conductive in less than a microsecond. The ions in the avalanche stream give up energy of ionization in the form of wideband rf energy and visible light. The rf energy is the source of RFI from the spark discharge.

Spark discharge seen on the tv screen appears as a band of horizontal dot-dash white spots, or "shot lines", moving slowly up the screen (Figure 5). The width and intensity of the lines are dependent upon the strength and severity of the interference. Tv viewers who watch signals in a fringe area often experience this interference from passing autos and trucks. A less-obvious source of spark interference is the common thermostat.

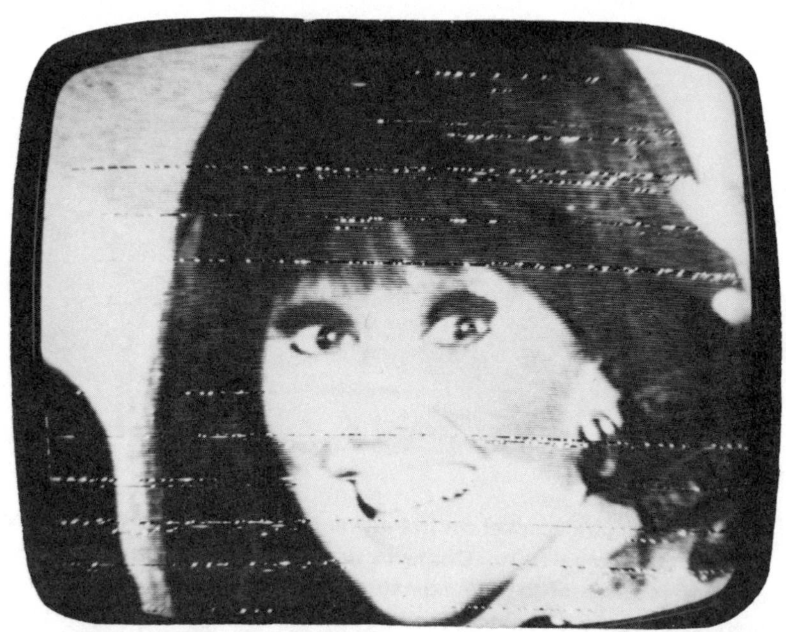

Fig. 5 (above) Spark discharge creates "shot lines" on TV screen and shows up as sharp pulses on oscilloscope pattern (below). This interference pattern is often caused by passing autos and trucks.

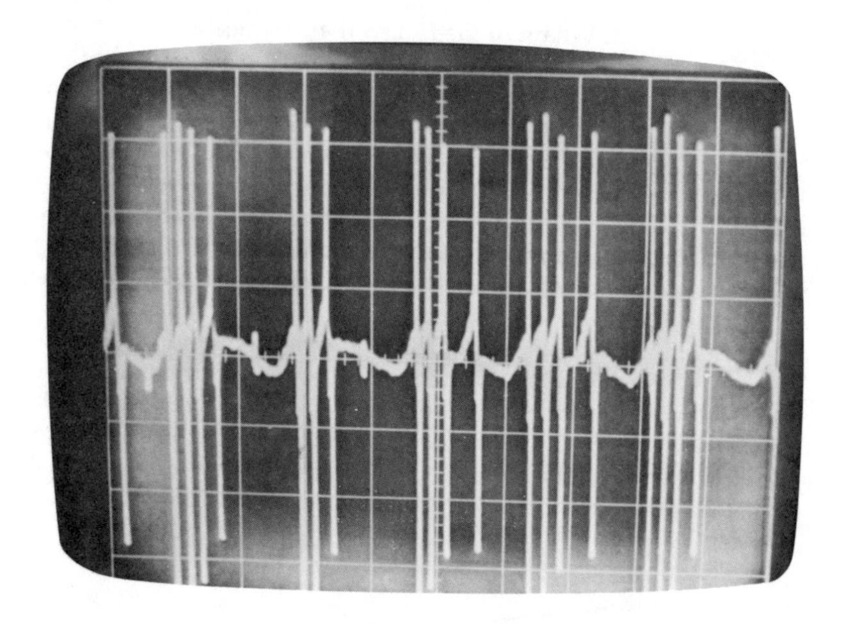

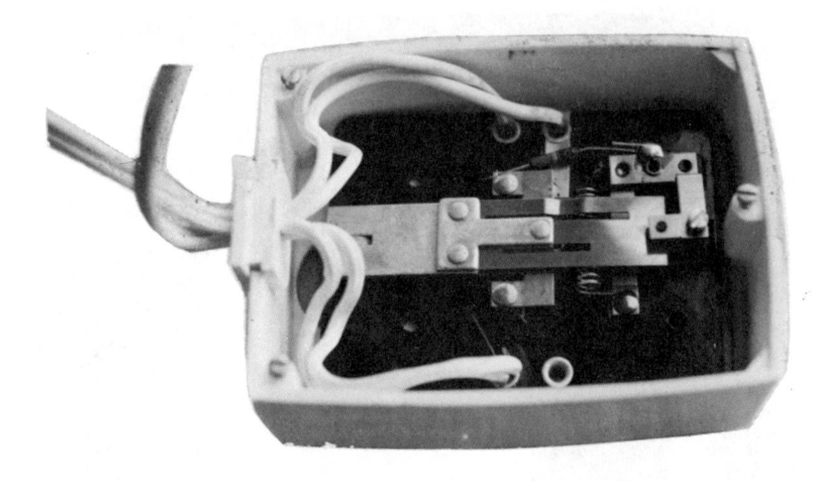

Fig. 6 An electric blanket control with snap-action thermostat that senses changes in temperature. Contacts are normally closed until heating element reaches desired temperature after which they snap open. As thermostat ages the contacts become pitted resulting in a buzz and chatter creating cyclic bursts of RFI. Placing a .01 uF, 1.4 kV ceramic capacitor across the contacts usually cleans up the noise. Old thermostats should be replaced.

A Villain in the House--the Thermostat

The thermostat is a temperature controlled electric switch made of two elements, a bimetallic strip that senses changes in temperature and a stationary contact (Figure 6). When connected to a load, such as a heating pad, electric blanket or aquarium heater the contacts are normally closed until the desired temperature of the heating element is reached, after which the contacts open, shutting off the electricity to the element. Some thermostats sense cold and turn on electrical equipment at a predetermined lower-than-normal temperature.

In addition to use in pads and blankets, the thermostat is also used in place of a fuse or circuit breaker on low voltage, low current devices such as a doorbell transformer and the contacts are normally closed. If a short occurs in the doorbell wiring, causing the transformer to overheat, the contacts open until the transformer cools. During this period the bell is inoperative. Not one homeowner in a thousand knows of the existence of this device.

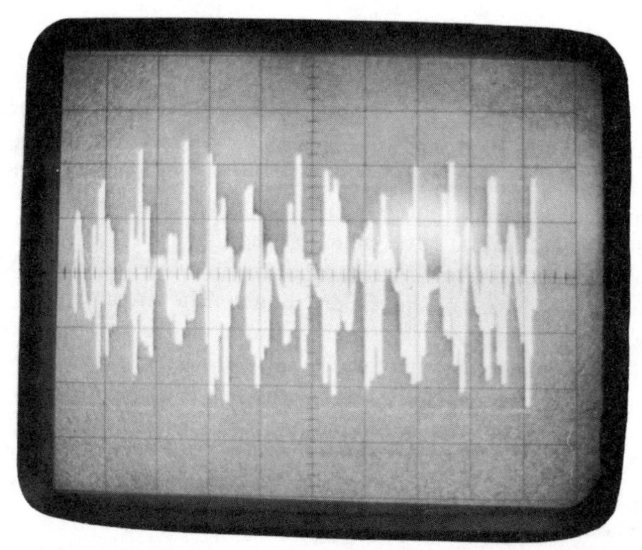

Fig. 7 Heating pad interference is representative of spark discharge but very heavy in intensity. Electric blanket interference looks much the same.

Newer doorbell transformers are made without a thermostat and when a short occurs a limiting device in the transformer burns out, opening the circuit.

In any thermostat, regardless of use, a spark of abnormal duration is generated between the contacts. This happens because the bimetallic strip in the device moves slowly. Over a period of time, repeated operation results in burning and pitting of the contacts. The pitting builds up a high electrical resistance in the form of metallic oxides which then cause the contacts to buzz and chatter. The result is a *bzzt, bzzt, bzzzzt* noise heard on a nearby radio and the appearance of "shot lines" on the tv screen.

Newer electric blankets and heating pads are equipped with a snap-action thermostat and the only RFI heard from this device is a quick "pop".

The cycling bursts of RFI from an older, defective thermostat vary from two to five seconds in duration, with off periods between the bursts of one to five seconds (Figure 6). In addition, some devices create a "popping" noise between the bursts. A few units will create a continuous buzzing noise without pause, but these devilish devices are few and far between.

The Heating Pad That Sent Morse Code!

The telephone almost jumped out of the investigator's hand. "You've got to help me!", pleaded the CB operator. "My neighbor has called her lawyer and she's going to sue me for being a public nuisance because of television interference. An I don't have any TVI on my own television set. It just can't be my station!"

True enough. The first investigation proved the CB set to be "clean", causing no TVI in the CBers home. The investigator called the complaining neighbor who said the interference was nearly continuous and so severe that some of the channels were impossible to watch. The neighbor also reported that adjoining neighbors were experiencing the same interference and it was undoubtedly due to the nasty CBer who "left his transmitter on the air whether he was home or not". The neighbor hotly maintained that the dot-dash lines seen on her tv screen proved the CBer was sending Morse code, too!

In a situation such as this the investigator hopes that the source of RFI will be in the residence of the complainant, especially if legal action is threatened. The investigator had a hunch that this might be so because of the severity of the interference to the complainant's tv set as compared to that of the neighbors.

The investigator drove to the residential area with his monitor receivers running and soon picked up the *bzzt-bzzzt-bzzzt* noise of spark discharge interference on the 20 meter ham band. He noticed it about two blocks from the complainant's house. As soon as he came within a half-block of the house he could hear it strongly on 6 meters. Finally, as he drove into the driveway of the residence, he heard it on his portable fm receiver. His hunch was confirmed: the source of the RFI was probably the home of the individual threatening the lawsuit. He went to the front door with his portable radio tuned near 88 MHz.

After introducing himself and explaining what he was doing, he mentioned to the woman who answered the door that he thought the RFI was generated by a thermostatically controlled device and suggested (very diplomatically) that it quite possibly was in her own home. After vehemently denying the possibility of creating her own interference, the woman reluctantly let the investigator into her house.

A child was sitting at the breakfast bar in the kitchen watching tv cartoons on one of the higher vhf channels. The woman switched to a lower channel over the protests of the viewer and the interference was immediately present: a bad case of spark discharge RFI.

Gently, the investigator questioned the woman. No, there was no butter-keeper in the refrigerator. No, she did not own a heating pad.

But yes! She remembered an old pad in the family room which she used about once a month when she had back pains. But it was turned off when not in use.

Following the woman down the hall into the family room, the investigator found the heating pad. The control was turned off, but the pad was suspiciously warm. Placing his ear to the pad, he heard the sparking noise of a defective thermostat. He pulled the plug, satisfied that he had found the source.

Returning to the kitchen, he was surprised to see that the interference still remained on the tv screen, although somewhat reduced. He immediately turned on his fm receiver and heard the spark noise once again.

The investigator looked around the kitchen and spotted a large fish aquarium on a corner table. One glance at it told him that the tank had a thermostat to sense the water, turning the heater on and off to maintain a constant temperature. From constant use, the contacts of the slow-break thermostat had become pitted and started to chatter.

There was only one cure: the thermostat had to be replaced with a noninterfering type. The investigator told the woman that a fish tank thermostat could be checked quickly for RFI by merely observing the small neon lamp that signifies the heater is on. If the lamp flickers, the thermostat is chattering and generating RFI and should be replaced.

Case dismissed!

Are You Your Butter's Keeper?

Complaint: Interference so bad that it was impossible to watch any television channel. Obviously the fault of the CB operator down the street! Neighbors having the same problem.

The RFI investigator quickly tuned in the characteristic noise of spark-type RFI on his car radio as he drove into the neighborhood. No, there were no thermostatically controlled devices in the home of the complainant. The investigator asked and received permission to trip the circuit breakers in the electric utility box near the kitchen door. The source of RFI was on the kitchen circuit. One by one the cords of the kitchen appliances were pulled, and the interference stopped when the refrigerator was disconnected. The problem was quickly traced to the butter keeper. The thermostat was bad. The cure? Disconnect the butter keeper or replace the thermostat.

What Did You Expect? Chimes?

The ham operator was getting spark-type interference, 15 seconds on and 15 seconds off, he told the RFI investigator. And the RFI was also affecting his tv set. Driving around the neighborhood, the investigator noticed the radio noise peaked at a nearby residence whose owner was in the process of writing a letter to the FCC about radio interference caused by the ham down the street. The investigator introduced himself and asked if he could listen to the racket on the radio. He carried a portable radio with him to check the noise. Interestingly, the noise peaked on his radio as he walked between the kitchen and the dining room. Above his head was the doorbell chime. He asked the home owner to go outside and trip the circuit breaker that supplied electricity to the kitchen area of the house. Sure enough, the noise vanished when the breaker was tripped. Upon asking about the chimes, the home owner said they were disconnected because his daughter often pushed the doorbell because she liked the sound of the chimes. Investigation revealed that the disconnected wire was touching the metal case of the chimes, thus inadvertently shorting the doorbell circuit through a high resistance to ground. The doorbell circuit, moreover, had been wired into the same power outlet as the television receiver, providing a ready conduit for the noise into the tv set, the noise then was further radiated to the neighbors' homes via the television antenna system. Insulating the bare end of the wire from the doorbell case cured the problem.

The Case of the Noisy Electric Blanket

Some years ago, when electric blankets were becoming popular, the investigator noted a bad buzzing noise in his receiver. Interference was also noted on the tv set. After a good deal of looking around, the investigator traced the noise source to his wife's new electric blanket. Again, it was the thermostat in the blanket control. The control was taken apart and the contacts cleaned. For a while the RFI disappeared but it soon returned, worse than before.

There was nothing wrong with the blanket except that it created RFI so it was finally given to a couple owning a camper who used it when they went on overnight skiing trips. Later experiments showed that placing a .01 uF, 1.4 kV ceramic capacitor across the contacts in the control box cleaned up the annoying interference.

Fig. 8 Electric "shocking" wire keeps Old Dobbin from nibbling the wood fence. Dirty insulators and branches touching the wire cause current leakage to ground and resultant RFI.

The Shoot-out at the OK Corral

Noise bursts of about half-second every second? The RFI investigator had never heard of this unusual noise pattern. But here was just such a complaint from an urban area. It was a puzzling problem, the investigator thought, as he drove around listening to the unusual noise on his car radio. It didn't seem to come from any of the homes. In fact, it seemed to come from the open fields which contained nothing but horses..

But wait! What was that wire running atop the fence on insulators? A close examination showed it to be some kind of electric wire (Figure 8). The complainant was contacted and he and the investigator walked out to the corral.

The owner of the horse explained that horses liked wood and the top bar of the fence was continually being nibbled away by the horses. To prevent this, a wire was strung along the top of the fence which would

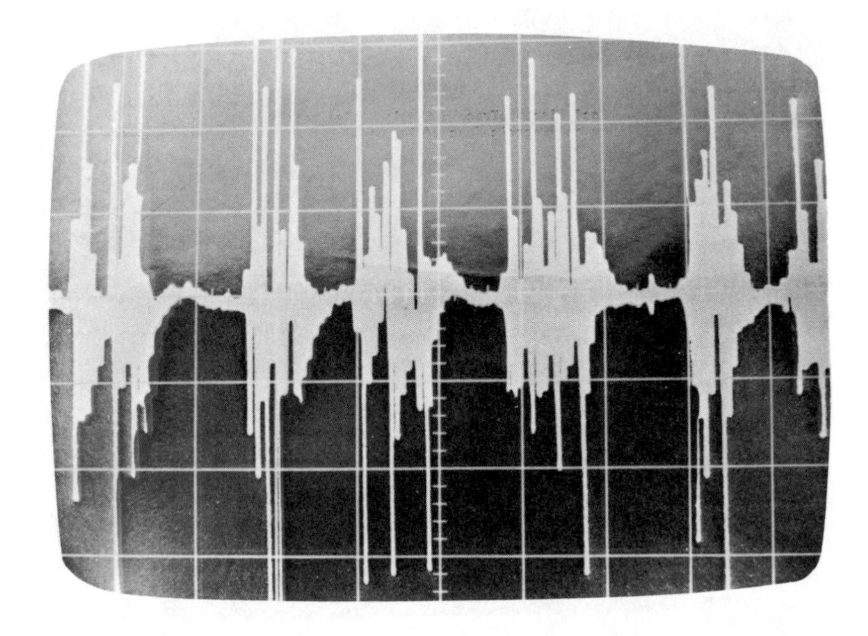

Fig. 9 Oscilloscope picture of electric fence interference reveals very high voltage pulses whose frequency can be varied by control box. RFI was eliminated by cleaning fence insulators and removing branches that touched the wire.

give the horse a mild shock each time he touched it. For maximum control of an obstinate horse, a small control box adjusted the intensity and frequency of the voltage which could produce up to 60 shocks per minute (Figure 9).

A close investigation of the shocking wire showed that the problem was a combination of difficulties. Dirty insulators and branches touching the wire caused a current leak to the ground at several points. In addition, there were poor splices in the fence wire and a badly rusted connection at a gate opening.

To determine if the trouble was in the fence charger or the wiring, the wires to the fence were disconnected, leaving the charger to run alone. All was quiet. If there was noise created by the charger, it could possibly be the pulsing unit or the charging capacitor. Happily, cleaning up the fence wire solved the problem.

Warning! If you have a problem with an electrified fence, do not touch the fence wire when the charger is operating. The shock is not harmful but it is powerful enough to give you a case of tight jaws!

Fence chargers can also be mounted on steel fences, and are sometimes found in the cities. Some city dwellers use an electric fence wire to keep large dogs in their backyards. So don't be surprised if the interference seems to come from outside a residence when logic tells you that the RFI-producing devices are inside the building.

Electric Clock RFI

RFI from a clock? Yes, it is possible. The complaint: intermittent television interference, not on all day nor every day. The RFI investigator noted a grinding, rasping noise in a small broadcast receiver near the tv set. It was on for 25 seconds and off for 35 seconds. For the better part of an hour the investigator drove around the neighborhood, driving for 25 seconds, then stopping for 35 seconds so he would not lose the "sense" of the racket. He was finally able to determine the dwelling from which the noise seemed to originate. After being admitted into the home, the investigator still found it difficult to pinpoint the source of the RFI. It was finally traced to an elegant old electric clock on the mantle over the fireplace.

It was noted there was a slight pause of the sweep second hand at the 35 second position at which point the noise started. The noise was heard until the second hand passed the 60 second position. The load on the motor created by the upward movement of the second hand created the RFI. When the old clock motor was replaced, the interference ceased.

Case closed!

Gas Appliance RFI--Spark Ignition

A gas appliance as a source of interference? Hard to believe. But most of them do use electricity. RFI from this source can be on for about 15 seconds, after which no noise is heard for up to 20 minutes: a most difficult noise source to locate!

It took the investigator three hours to locate his first case of gas igniter RFI caused by a gas operated clothes dryer. The igniter was a crude spark gap activated for about 15 seconds until the heat of the burning gas opened a thermostat, shutting off the spark. The igniter would remain off until the internal temperature of the dryer droppped, at which time the igniter would fire the gas again. When the igniter was replaced with a new one, the interference disappeared.

Spark igniters are often used on gas ovens in place of the old fashioned, energy wasting pilot light. Top burners in many gas ovens use

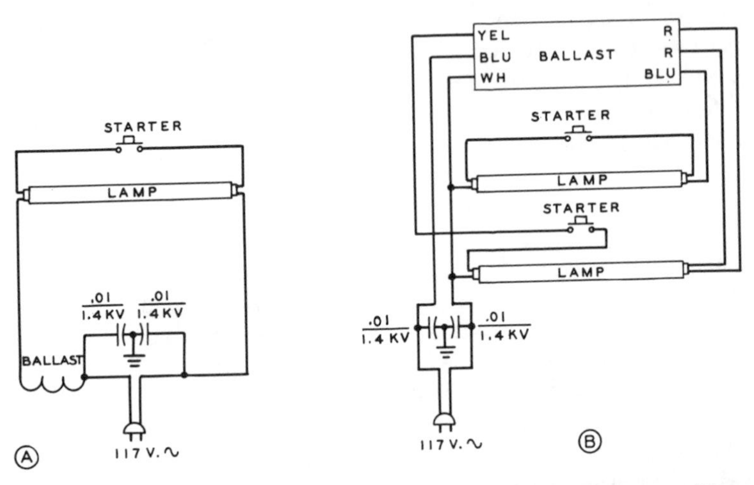

Fig. 10 Various types of fluorescent lamps and ballast systems are in use. In all cases bypassing both sides of the line to the lamp frame will reduce radio noise. Keep lamp away from radio receiver to reduce direct noise pickup.

one spark igniter for each pair of burners. The spark occurs between the igniter and a ground dimple on the oven surface. Replacement of the igniter usually clears up this interference.

The Case of the Noisy Fluorescent Lamp

The common fluorescent tube lamp is an electric discharge device wherein the mercury vapor in the glass tube is acted upon by a stream of electrons which activates the fluorescent materials inside the tube (Figure10). The phosphors transform the radiant energy from the discharge into visible light.

The fluorescent lamp, operating on the 60 Hz household current, is turned on and off 120 times a second. This quick action is not visible to the eye but the combination of the electron stream and the rapid interruption of the energy gives rise to rough radio waves that can cause severe RFI. The radio interference generated by the lamp electrodes may. reach a nearby receiver or entertainment device by direct radiation from the lamp to the radio antenna or by radiation back along the electric line to the receiving equipment.

A fluorescent lamp rarely interferes with a television receiver as the noise frequencies generated by the lamp generally fall in or near the standard brodcast band (540-1600 kHz), with a secondary peak in the 7-9

MHz region; the latter frequency peak is dependent upon lamp size. Additional noise can be generated in the region of the 20 meter band. The noise is coarse and guttural and continuous while the lamp is lit.

In addition to the distinctive RFI noise, the fluorescent lamp has an interesting peculiarity. For some obscure reason, the interference created by a number of lamps is not additive. That is, if there are ten fluorescent lamps creating RFI, you will generally not hear a total of all the noise produced, you will hear only the lamp creating the greatest amount of RFI. When that lamp is turned off, the next loudest one will be heard, and so on.

Generally speaking, the RFI directly radiated from a flourscent lamp is dissipated within a few feet of the lamp and can be controlled to a great extent by proper positioning of the lamp and the radio. Small lamps (up to 32 watts) require a minimum separation of 5 feet between lamp and radio. Lamps up to 100 watts capacity require a separation distance of about 10 feet.

In case the radio or electronic equipment cannot be moved clear of the radiation field of the lamp, the antenna of the receiver should be moved away from the lamp. The modern solid state home radio, however, is commonly equipped with a built-in antenna which presents a difficult problem. Some radios have provision for an external antenna and ground connection and the use of these will help to reduce lamp noise.

Radio interference filters for fluorescent lamps are available on special order from the lamp distributor (Figure 10). A filter should be installed on the lamp fixture as close to the lamp terminals as possible.

The Noisy Neon Sign Mystery

A neon sign is a glass discharge tube containing neon gas which, when ionized by a high voltage, will glow (Figure 11). The color of the sign is determined by the gas used. Neon provides a red glow and other gaseous elements provide other colors. High voltage is required to operate a neon sign and the sign is a prolific source of RFI. Some of the noise is caused by the high voltage sparking to nearby metal objects or by loose electrical connections, but most of the noise is caused by loss of gas pressure in the sign which causes flickering. The on-off ionization of the neon gas causes radiation of a rough, spark-like radio disturbance that can travel for a great distance.

Modern neon signs have radio noise filters installed but older signs usually do not. There is not much one can do about neon sign interference except to ask for cooperation from the owner who can contact a sign

Fig. 9 Modern neon sign has built-in RFI suppression circuits. Older signs create severe interference and a sign company should be contacted for repair or maintenance of the offending device. Rf filters are available, but they should be installed by one familiar with the operation of the sign.

company for maintenance or repair of the offending unit. RFI filters are available for neon signs, but they should be installed by one familiar with the operation of such a sign.

The RFI investigator discovered bad interference on his own ham set one day. With the aid of his rotary beam, he found the general direction of the noise and tracked it to a neon sign more than two miles distant. It was an animated sign over a nightclub. First, the letters TOP would flash, followed by the letters LESS, and then the complete word TOPLESS would flash.

The investigator told the owner of the sign that it was causing severe radio interference, but the owner could care less. He told the investigator that he was probably the ham they were always hearing on their stereo music system whenever the topless dancers were at their best. The sign is still flashing but, luckily, the investigator was planning to move away.

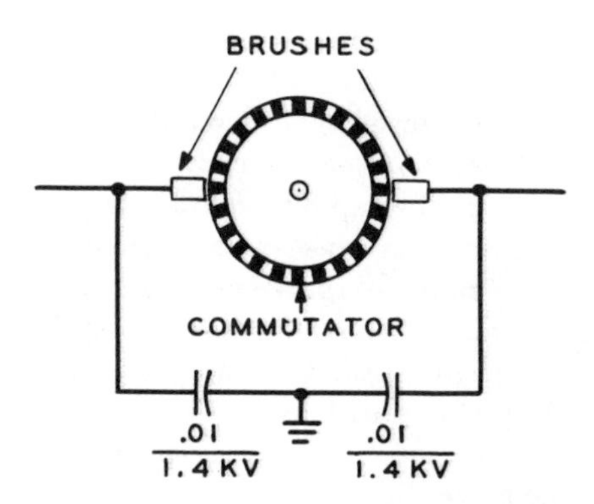

Fig. 12 Brush noise of small electric motor is eliminated by bypassing each brush to metal frame of the motor. If inconvenient to do this, a single capacitor placed across the line near the motor will help. As a last resort, a line filter may be placed between line cord and wall plug.

The Case of the Noisy Food Blender

Wow! It nearly blew the headphones off the ham operator: A noisy whine appeared for a few minutes and then vanished, only to start up again, wiping out the 40 meter band.

Removing his headphones, the amateur was surprised to hear the same noise coming from the kitchen. It sounded like his wife using her food blender.

And that is what is was. The blanketing racket came from the motor of the blender. Looking through the ventilation louvres covering the motor, the amateur saw the brushes of the little motor sparking vividly. As soon as the cake was completed and the blender cleaned, the amateur took it into his shop and installed bypass capacitors from each brush to the metal frome of the motor. Small .01 uF, 1.4 kV disc ceramic capacitors were used (Figure 12). And sure enough! The motor noise was eliminated except for a very weak whine that could still be heard in the kitchen radio when it was tuned off-channel. The noise was completely cleaned up on the ham bands.

He told the story to his friend the RFI investigator the next day. The investigator pointed out that plug-in type line filters are available from several manufacturers that suppress interference from brush-type motors.

Cleaning up Spark Discharge Interference

Spark interference is caused by the irregular flow of current between two points separated by air. A spark is created when the voltage is high enough to ionize the air in the gap. The spark sets up an electromagnetic radiation field about it which causes interference in a nearby radio or television receiver. The radiated spark energy reaches the receiver either directly through the air or by conduction along the power cord and electrical wiring in the building. In the majority of cases, the interference can be suppressed or eliminated by the use of a *noise filter* which suppresses the energy in the spark and prevents it from being radiated or conducted down the power line.

The Capacitive Filter - Special Capacitors Used

The simplest interference suppression device is a small capacitor placed at the terminals of the spark discharge (Figure 12). The capacitor may be placed across the sparking terminals or two capacitors may be used to bypass both sides of the gap to the metal frame of the device. Both methods work, but often one is more effective than the other.

Special disc ceramic capacitors are available for this job and their use is recommended. The "garden variety" 600 volt disc capacitor is not recommended as it is not tested for use when a continuous ac voltage is applied across the terminals. Various manufacturers market special capacitors for this service that are rated for continuous operation at 125 volts ac and 1400 volts dc. (The dc rating is required because of high voltage transients that often appear on the power line). The capacitors are tested at 2800 volts dc.

Three well-known brands of .01 uF disc capacitors designed for line filter service are:

Aerovox type AC-7
Centralab type CI-103
Sprague 125L-S10

These capacitors, or their equivalents, are recommended for use as interference suppressors for the 120 volt, 60 Hz power line.

The Line Filter

If it is not feasible to mount a capacitor directly at the sparking terminals of the device or if the capacitor doesn't suppress all the RFI, it

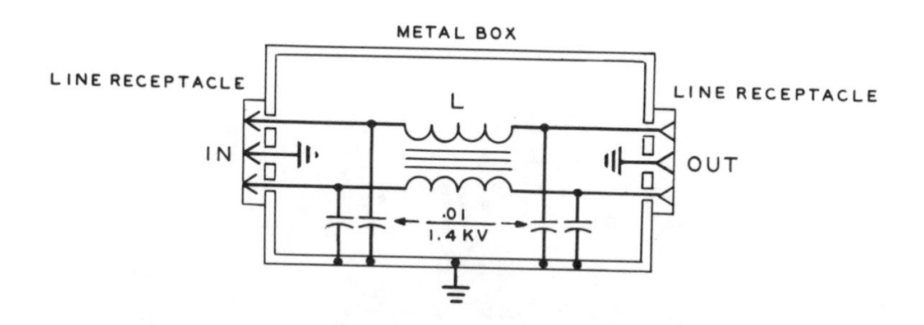

Fig. 13 Simple line filter you can make. Dual winding coil is wound on ferrite rod (J.W. Miller FR-500-7.50 or Amidon R61-050-750). The rod is 61 nickle-zinc material having a permeability of 125. The wires are laid in parallel and wound on core at the same time. Ends are held in place with twine and epoxy cement. Filter is built on a metal plate bolted within an aluminum chassis which serves as a dust cover.

is usually possible to use a line filter (Figure 13). The line filter is made up of bypass capacitors which provide a low impedance to ground for the noise signal and series inductors which provide a high impedance path in series with the line for the noise. The filter is built within a metal box which is grounded to the case of the device. Special filters for this application are made by several manufacturers. These filters are rated for a maximum value of line current (determined by the wire size in the coils) and are commonly available in ratings of up to 20 amperes and up to 100 amperes on special order.

A Line Filter You Can Build

Construction of a line filter is shown in Figure 14. The unit is built in a small metal chassis that has a bottom plate held in position with sheet metal screws. Male and female line connectors are mounted on the ends of the box. In some instances, the connectors may be omitted and the filter connected directly in the line.

The filter coils are wound of insulated wire on small ferrite rods which provide a maximum value of inductance for a given number of turns. One coil is placed in series with each side of the power line. The filter capacitors are mounted on small phenolic tie-point strips which

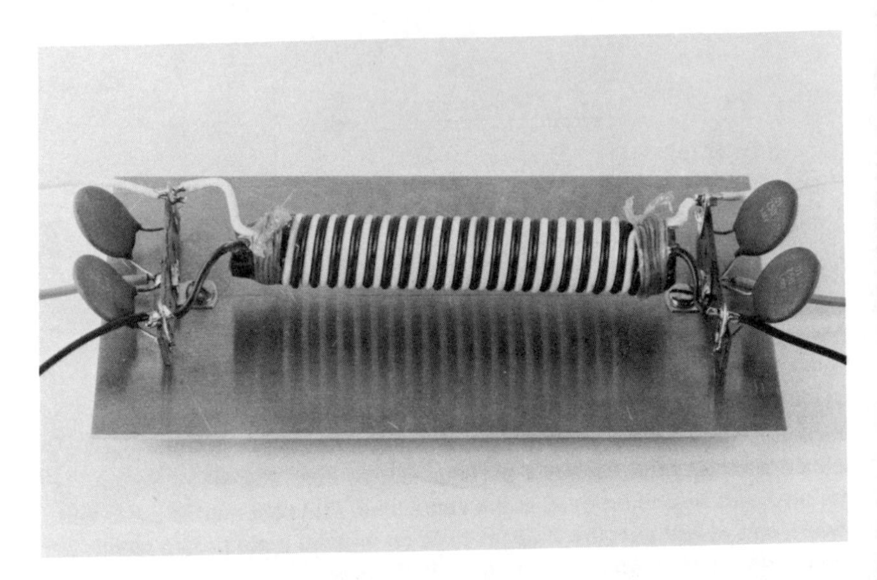

Fig. 14 Heavy duty line filter. Inductor consists of two coils, each 20 turns of 18 gauge insulated wire wound on ferrite rod 1/2-inch diameter and 4 inches long (see Figure 13). Rod is trimmed to length by notching it around the circumference with the sharp edge of a file and breaking it at the notch over the edge of a table. Windings are held in place by string wrappings at each end which are covered with epoxy cement. Capacitors are 1.4 kV working voltage disc ceramic units (see page 40).

also support the ends of the coils. For use with devices drawing up to 600 watts from the 120 volt line, no. 18 gauge wire can be used for the coils.

The filter should be connected to the metal frame of the noisy device. If it cannot be mounted on the device, it may be inserted in the line with the case unattached, however, the effectiveness of the filter will be reduced as the line cord between the filter and the device radiates interference.

Whenever the use of such a filter is contemplated, the device should be inspected to make sure it is not malfunctioning. Severe cases of RFI caused by electric motors has been cured by merely replacing badly worn motor brushes. So make sure an offending motor or other electrical device is in good operating order before you attempt noise suppression of any kind, or you might be wasting your time and fraying your nervous system.

Chapter 3

Electrostatic Discharge--ESD
A Little-suspected Source of RFI

A mysterious cause of RFI is *Electrostatic Discharge* (ESD) which implies the presence of static electricity, or the accumulation of electric charges on a surface. Everybody is familiar with the suprising shock received after walking across a dry rug and touching a doorknob in clear, cool weather. On a much larger scale, static charge concentrations in thunderclouds can produce potential differences between clouds and ground of millions of volts (Figure 1). A breakdown between charge centers in clouds, or between clouds and earth, results in a lightning discharge. A nearby lightning stroke can disrupt radio communication and ruin tv reception because of the electromagnetic energy liberated during the flash.

On a smaller scale, static electricity depends upon an excess of electric charges with respect to ground, or charges of the opposite polarity. Generation of charges (commonly known as static electricity) usually results from a mechanical process. Well-insulated metallic objects can be charged when brought in contact with and then separated from insulating materials.

An antenna in moving air which contains dust particles, water or snow can become charged with static electricity; the potential reached may be several thousand volts. It is limited only when corona discharge occurs from a point of high electrical stress. In the old days of sailing ships, static electricity resulting from charges collected in the sails during a heavy storm manifested itself as a corona discharge called "Saint Elmo's Fire" by superstituous seamen (Figure 2).

Fig. 1 Lightning flash from earth to ground represents a great display of energy. The studies of lightning indicate the flash is very complex, being composed of many separate strokes from the positive potential of the cloud to the negative potential of the earth. A brush discharge from points on the earth may trigger a massive flash. Clouds rapidly rebuild electric charge in a few seconds after lightning flash.

Electromagnetic radiation from a lighting discharge can propagate over a long distance and is easily reflected by the ionosphere. Natural lightning discharges in the tropic zone create radio static which covers the globe. Man-made discharges, although weaker, can disrupt electromagnetic communication over a wide area.

The Electrostatic Discharge and RFI

Electrostatic discharge can be created by two or more metallic objects making intermittent contact. And in addition, an electric discharge can be induced in a metallic object which is in the field of a strong electric power source. The sound of ESD in a receiver is similar to the static heard

Fig. 2 Corona discharge known as "Saint Elmo's Fire" terrorized seamen on early sailing ships. The discharge takes place around masts, spars and other elevated structures on wooden ships in an electrical storm.

Superstitious sailors regarded corona as an omen of evil. Christopher Columbus notices "celestial lights" on the masts of his ship in 1453. (Drawing and data from "Thunder and Lightning", by de Fonvielle, Charles Scribner and Sons, New York, 1869).

during a lightning storm-rasping bursts of noise which may subside to crackling, scratching noise. On the tv screen, the ESD interference shows up as horizontal dot-dash white spots, similar to the image of spark discharge. In most cases of man-made ESD interference, the electrostatic discharge source is quite close to the point of interference and is often caused by two pieces of metal rubbing together. Because of the subtle nature of the cause, ESD is often extremely difficult to locate.

Several clues help the investigator to locate a source of ESD:

ESD often occurs during windy days
ESD often occurs during dry days of low atmospheric humidity
ESD often occurs when the temperature changes
ESD usually occurs when objects are moved or vibrated

The following true life histories illustrate typical cases of ESD interference observed and cured by the RFI investigator.

The Case of the Irate TV Viewer

It was all the fault of the neighborhood radio ham, of course. The complainant had "shot lines" and "snow" all over his tv screen. Time of occurence: when the afternoon breeze came up.

The RFI investigator drove to the address and heard the sound of ESD on his car receiver as he approached the house. He took his portable radio and a battery-powered tv set into the complainant's house to observe the noise. Yes, there was the same characteristic interference on both of the investigator's receivers as well as on the tv set in the home.

As an experiment, the owner's tv was shut off and the plug pulled from the wall. The ESD immediately stopped on the investigator's equipment. Going outside, the investigator checked the roof-mounted tv antenna with his binoculars and observed that the connections between the antenna and the ribbon lead-in line appeared to be badly corroded. The home owner climbed a ladder to the roof and confirmed that the connections were loose. He attempted to tighten the wing-nut on one antenna terminal and in doing so the lead-in wire came apart in his hands.

The owner admitted the tv antenna had been up for about fifteen years and the investigator suggested that a new antenna and lead-in would probably clean up the problem. A follow-up phone call confirmed this.

Question: When was the last time you inspected your tv antenna and lead-in for rusty, half-broken joints and loose connections?

The Mystery of the Unhappy CB Salesman

On a blustery spring day the RFI investigator responded to an interference call from a large CB dealer. The salesman told the investigator that he was losing CB sales because of bad power line noise that blocked out reception on demonstrator sets. Why didn't the power company instantly fix their noisy, defective power lines?

The investigator detemined that the disruptive racket was some sort of electromagnetic discharge as it seemed to occur only on dry, hot, windy days. On some days the noise was so bad that the local CBers were circulating a petition demanding that the power company do something about the racket which extended for some distance from the CB store.

The RFI investigator asked the salesman to turn on a demonstrator set. Sure enough, an intermittent, raspy noise was heard, loud enough to drown out most CB signals. After looking around the premises, he went outdoors. The first thing he did was to shake the metal mast holding up the dealer's CB antenna. The salesman rushed out of the store, demanding to know what was going on as the radio interference has increased a thousandfold!

The investigator looked up and saw that the antenna mast brushed against the metal gravel stop of the roof. He also noticed that a pipe strap braced the mast to the side of the metal building. Holding the mast firmly against the breeze stopped the noise completely.

He told the shop owner that the problem was caused by intermittent metallic contact between the mast and the metal gravel stop. The investigator explained that the interference occurring during hot, windy days was caused by minute, charged dust particles blowing against the mast and creating a static charge on the metal surfaces. When the static charge was great enough it would spark to a surface having a lower potential, such as the gutter or metal building, both of which were grounded. As long as the surfaces were in contact, the discharge took place peacefully. An intermittent contact, however, caused sparking which produced the loud radio noise. The shop owner was told that electrostatic potentials of more than a thousand volts could be built up on an insulated metallic surface when a dry wind was blowing, just as static electricity builds up on a person when he shuffles across a rug on a dry day. In this fashion, blowing dust creates a high potential and static discharge between nearby metallic surfaces, as the CB store owner found to his sorrow. The solution was to bolt everything together so that the static discharge was harmlessly conducted to ground.

The Case of the Induced ESD

Complaints of radio and tv interference were received by a local power company from customers whose homes backed up against a high voltage power line. The interference was noticed on warm, dry, windy days. Would the power company please repair its power lines and clean up the interference? Sensing an ESD problem ahead of him, the RFI investigator arrived on the scene and identified a noise he took to be ESD. He drove slowly along the power line listening to the noise on the car receivers, looking for a possible source of trouble. Nothing. But in one area he noticed that a movable, corrugated iron building mounted on skids was located near the power line. It was a temporary workman's shack, left from some forgotten project. The radio noise was particularly loud in the vicinity of the little building.

The investigator got out of his car and walked toward the building. It was empty and deserted. As he approached the radio noise in his hand-held receiver grew in intensity. The noise must be coming from the building.

He walked around the shack several times but it was not until a gust of wind vibrated one of the building's corrugated metal wall panels that he located the noise source. The various panels were loose and held in position only by wobbly nails. Each time a panel moved a fraction of an inch an electrostatic discharge would take place to an adjoining panel. The induced charges came from the nearby power lines which had a strong electric field about them. The investigator made a temporary fix by pounding the nails back in place and the radio noise stopped.

The noise had been caused by induced static voltages from the high tension lines. When the panels moved a discharge was evident, creating radio noise which was induced back into the power lines and transmitted up and down the conductors for many miles.

The building owner was finally located and asked to remove the building, thus eliminating the source of future problems. Case solved!

The Mystery of the Baffled Radio Ham

The RFI investigator received a frantic call from a local radio ham. "Come right over. I have terrible power line interference!"

Responding to the plea, the investigator found the ham lived close to a large power line made up of two 220 kV and two 66 kV circuits. The ham's beam antenna was only about fifty feet from the lines. The only time the

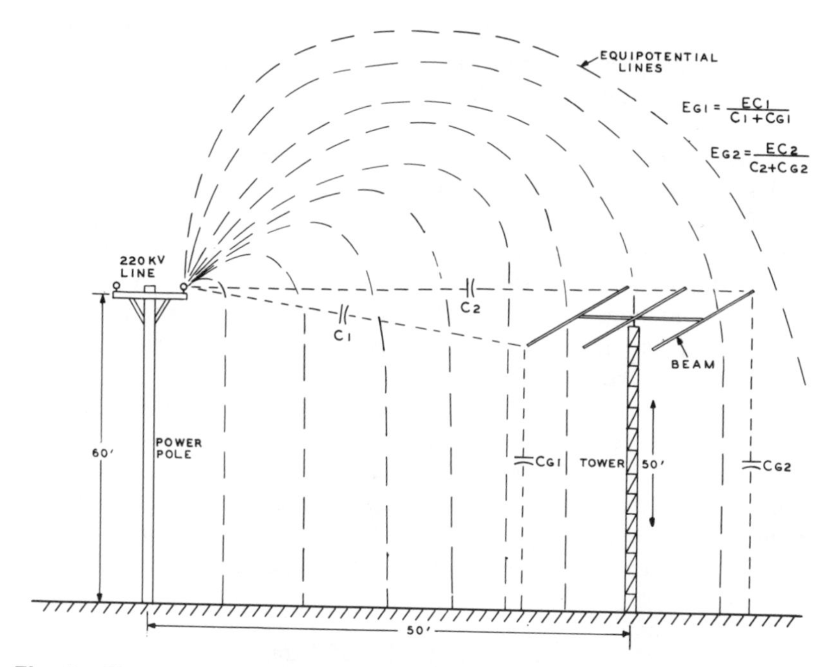

Fig. 3 Strong electric field around high voltage line induces voltages into tower hardware and antenna. Induced voltage is proportional to capacitances between structure and power line and structure and ground. Because of high impedances involved, potentials of the order of several hundred volts can be developed in spite of the small capacitances.

poor fellow could operate without intolerable interference was during the rainy season when, presumably, the rain washed off dirty insulators on the power lines. The ham had been off the air ever since he erected his new tower and the noise seemed to get worse with each passing day.

The investigator heard the noisy racket on the ham's station receiver but noted that he could not hear the noise on his own car receiver parked in front of the amateur's residence. The noise seemed to be concentrated around the antenna tower.

Convinced this was an ESD problem, the investigator donned a body belt, safety straps and heavy rubber gloves and climbed the amateur's tower. He found the tower guys loosely attached to the metal tower and with a voltmeter he measured nearly 750 volts between the tower and the guy wires! He also found induced voltage between the beam and the tower, and between the loose-fitting sections of the tower.

Knowing nothing about the induction capabilities of high voltage circuits, the amateur had installed his beam and tower much too close to the 220 kV circuits (Figure 3). While the installation was far enough away from the lines for safety, the strong electric field surrounding the lines induced static voltages in the tower hardware and beam antenna.

The investigator recommended that the amateur move his tower at least 80 feet away from the power line, fasten the guy wires securely to the tower, install low resistance jumpers across the tower sections and ground the base of the tower.

Some time later, a follow-up phone call revealed that the tower had been moved and the ESD had disappeared. The problem of induced voltage was solved.

Remember, any loose connection in a strong electric field can produce dangerous voltage and interference. Power company safety rules require that a dead line being worked on in proximity to an energized line must be grounded to drain off any induced charge. The investigator can attest to the times he has been "bitten" by the induced voltage from a nearby line when he was working on poles as a lineman before this rule was put in effect.

An Expensive Bit of Wire

It took three months and the efforts of four RFI investigators to locate and correct a source of ESD that was affecting over 300 residents within a three mile radius of the source, plus several radio amateurs and CBers living up to six miles distant.

The problem seemed to be centered around a power line carrying 220 kV from a generating plant to a substation. The line was on metal towers placed on hill tops, the wire spans running almost 1500 feet long. Above the canyons, the lines were nearly 300 feet above the ground. The noise seemed to peak at one particular tower; insulators and hardware were inspected and replaced but the noise continued. De-energizing the line for a second or two stopped the racket but the cause of the noise could not be pin-pointed by the team of investigators.

Finally, one of the investigators, using a pair of high power binoculars, checked the tower and conductors from the ground. Passing his vision along the overhead cables he noticed an object hanging from one of the wires near the center of a long span over a canyon. It looked like a piece of wire hanging from a conductor. He lowered his glasses to rest his arms and neck, then looking up again he could not locate the wire.

The investigator called for a patrol crew and explained his findings. The crew climbed the tower and threw a long cotton line over the conductor. A wooden hook was attached to the end of the line. The hook was pulled up to catch on the line and then pulled along the conductor. Out over the middle of the canyon, the hook caught something and the mysterious object fell to the ground. It was a piece of copper wire about three feet long, bent into a U-shape. The interference stopped the moment the wire was removed from the conductor.

How had the wire been hooked over the conductor nearly at mid-point of a 1200 foot span over a canyon? One guess was that a large bird, thinking it had found something for its nest, was carrying the wire and dropped it on the power line. But who knows?

Initially, the copper wire had made good contact to the power line, but as the wire corroded contact became intermittent and an electromagnetic discharge took place between the wire and the high voltage conductor. The radio noise was then conducted back along the power line.

A condition such as this rarely occurs on high voltage transmission lines but does occur on low voltage distribution circuits, especially in areas where newspapers are delivered in wire bundles to paper boys. For some reason, the paper boys prefer to throw the wires in the air rather than dropping them in trash barrels and it is considered quite an accomplishment to hook the wire over a nearby power line! Paper boys are not always to blame for incidental wires on power conductors, sometimes line crews fail to take off the tag wires on transformers and conductors leaving a potential source of ESD for a future day.

The Triboelectric Table

Static electricity (ESD) results when two different materials come in contact or rub against each other and then are separated. The magnitude and polarity of these static charges depend upon the relative position of the materials in the *triboelectric table.* Materials above the reference point (cotton) tend to lose surface electrons when rubbed and thus acquire a positive charge. Those below the reference point tend to acquire a negative charge. The farther apart the materials are on the table, the greater the resulting potential.

Good insulators hold a static charge for a long time and high voltages can be built up under suprising conditions. Unrolling a few inches of Scotch tape can create a charge between layers of tape as high as 4,000 volts. Lifting a wool blanket off a polyester or orlon sheet or cover can create static charges two to three times this value.

TRIBOELECTRIC TABLE

Air
Human hands
Cat fur
Glass
Nylon
Wool
Silk
Aluminum
Paper
Cotton (Reference Point)
Hard rubber
Brass or silver
Rayon
Polyester
Orlon
Saran
Polyethylene
Polyvinyl Chloride (PVC)
Silicon
Teflon

Static electricity can damage various types of transistors and ICs as well as small thin-film resistors commonly used on printed circuit boards. Solid state equipment can sometimes be rendered inoperative merely by walking across a static-prone rug and touching a panel control.

Certain solid state devices, such as the newer CMOS series, are relatively secure against static discharge as they are protected by built-in diodes which short circuit unwanted static current. Even though an ESD pulse may have quite high voltage, the energy content is low and the diodes provide adequate protection.

A useful product that helps reduce static electricity is an antistatic liquid which forms a high resistance coating on an object that dissipates any charge buildup. One well-known product is *Downy Fabric Softener.* Other commercial products are *Staticide,* made by Analytical Chemical Laboratories, 2424 Pan Am Blvd., Elk Grove, IL 60007 and *79 Concentrate* made by Merix Chemical, Inc., 2234 East 75th St., Chicago, IL 60649.

An electrostatic meter that registers the ESD field is obtainable from various manufacturers including Static, Inc., Box 414, Lee, MA 01238; Richmond Division of Dixco Co., Box 1129, Redlands, CA 92373 and Simco, Inc., 920 Walnut St., Landsdale, PA 19446.

Chapter 4

The RFI Investigator

Tracing and Locating RFI In The Home

The previous chapters discuss some of the many sources of RFI and showed how a trained investigator can locate these trouble spots. While some municipalities and power utilities have their own RFI investigators, the majority do not. Even so, it is possible for the radio amateur or CB operator to trace sources of noise and interference for themselves and as a service to their neighbors. This chapter tells you how to do this, and the radio equipment you will require to be an RFI investigator.

Investigating the Radio Noise Spectrum

It is helpful to summarize three important points made so far:

1- Radio noise is distributed unevenly over the radio spectrum (Figure 1). For a given source, the noise level usually is greatest at the lower frequencies, diminishing in strength at the higher frequencies. Above 100 MHz, man-made noise is quite weak and RFI rapidly becomes weaker as the frequency of reception increases. As a result, the higher the frequency that you can observe the noise, the closer you are to the source.

2- Armed with a portable multiband receiver, the investigator notes the highest frequency on which the noise is heard. If the noise is noted in the broadcast band, or on the very low frequencies, the noise source may be quite a distance away. If the noise is heard in the shortwave spectrum as well, the source must be much nearer. And if the noise can be heard in the fm broadcast band, or the amateur two meter band, the source must be very close.

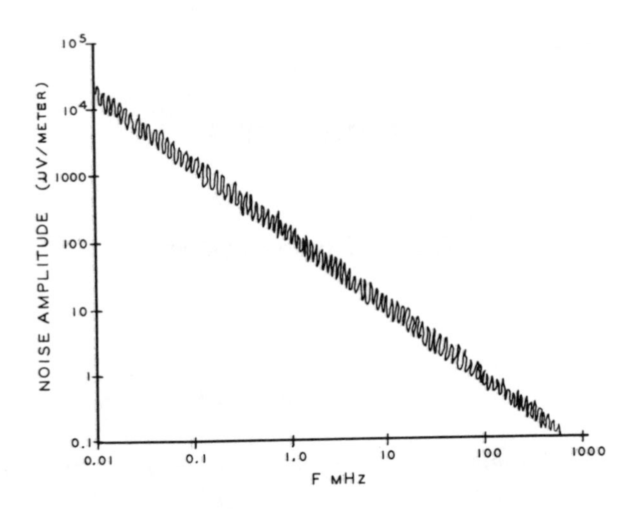

Fig. 1 The radio noise is distributed unevenly over the spectrum, the level being greatest at the lower frequencies and diminishing in strength at the higher frequencies. Above 100 MHz man-made noise is quite weak. Thus, the higher the frequency on which you can observe the noise, the closer you are to the source.

3- The instrument to conduct a noise search is a battery powered, portable am-fm receiver having at least one shortwave band. One that includes one or more public service bands is excellent for this type of work. Imported receivers costing less than one hundred dollars are available that will do the job.

Investigating and Locating Spark Discharge RFI

Interference to radio and television receivers can be caused by spark discharge from other home electrical devices. The white "shot lines" on the tv receiver (Figure 2) and the staccato buzz in the radio receiver offer clues to this type of interference. (Noise is seldom heard in the tv set because the audio portion of the program is transmitted and received by interference-free frequency modulation).

It might be thought that television interference could be easily tracked by using a small, battery powered tv set. Experiments have proven the idea to be impractical; the portable radio is a much better device for tracking RFI.

Fig. 2 Horizontal, white "shot lines" on tv screen and the staccato buzz in the radio receiver offer clues to spark discharge interference.

If the unwanted interference can be seen on a television receiver in a complainant's home, a note should be made of the highest channel on which the interference is present. If the interference can be seen as high as channel 7, for example, the source of the interference is nearby. If it is only noticed on the lower tv channels, it indicates the source may be some distance from the receiver. (Vhf tv channels 2-6 use 54-88 MHz; channels 7-13 use 174-216 MHz and uhf tv channels 14-83 use 470-890 MHz).

Once the investigator has studied the characteristics of the interference as it appears on the tv screen, he should turn on his portable radio to determine the highest frequency on which the noise can be heard. A striking similarity between audible radio noise bursts and visual interference is commonly noticed. It may be necessary to move the radio away from the television receiver to accomplish this test as most tv sets are prolific generators of radio noise themselves and you don't want the tv set radiation to block out noise reception.

Fig. 3 Radio noise is traced with a portable battery operated multiband receiver. Built-in receiver loop has directional response with respect to noise in broadcast band. When tracing noise on high frequency bands antenna of receiver is extended and held horizontal with respect to ground. Final search is made on highest band on which the noise is heard.

Check Your Own Residence First

The first RFI suspect is your own home. Before you go racing down the street, radio in hand, make sure your own residence is RFI-free. (One small municipally-owned power utility cut its cost of RFI-tracing by mailing out a questionnaire and instructions to customers complaining of RFI. It told them how to check their own home for interference. Thirty percent of those complaining found the source of RFI in their own residences).

The easiest, fastest way to check your residence for RFI is to go to the main fuse panel or multibreaker distribution box while listening to the noise on a portable receiver and de-energize each of the house circuits, one by one (Figure 7). If the RFI stops when a circuit is opened, you need

Fig. 4 Teen-age RFI hunter with portable radio is unnoticed whereas adult would attract attention of inquisitive neighbors. Loop antenna of portable radio provides excellent directive "fix" on local noise.

look no further. It's now a case of determining which part of the house is served by that particular circuit. Energize the circuit and start your search in that area by disconnecting appliances and lights one at a time. Chances are that you will locate the culprit when the noise stops as the defective device is disconnected from the line. If the RFI persists after you have opened all the circuits in your residence, you can be reasonably sure the noise source is elsewhere.

On the Trail of the Noise Source

Once the noise has been identified on the portable receiver and the home residence given a clean bill of health, the investigator is ready to trace the source of the noise outside the house. If the complaint is from a neighbor, he should walk around the neighbor's house, checking the noise level. He then travels slowly up and down the street holding the portable radio and rotating it about until the noise source is the loudest. He notes that the portable radio has a directional response with respect

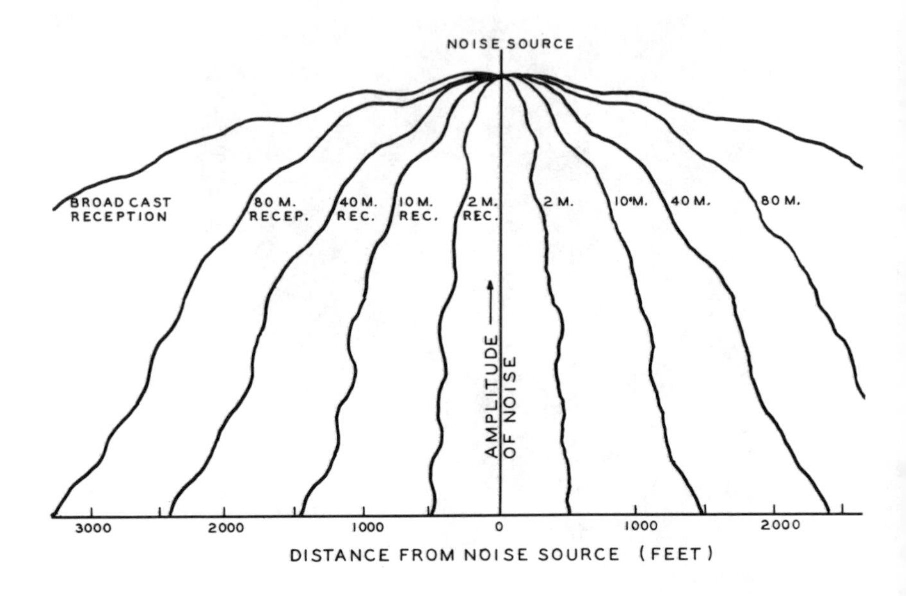

Fig. 5 The lower the frequency of noise reception, the greater the distance to the noise source. Listen on the highest frequency at which you can hear the noise, since you will be in the smallest "noise zone".

to the noise. This is due to the built-in loop antenna in the small receiver. Loudest noise reception is usually achieved with the case of the receiver in line with the noise source (Figure 4).

As the noise source is approached, receiver noise becomes louder. When the noise is very loud on the broadcast and shortwave band, the receiver should be switched to the fm band, or the highest band in which the noise can be heard. It is important to remember that the lower the frequency of noise reception, the greater the distance to the noise source. Thus, it pays dividends in time and effort to listen on the highest frequency at which you can hear the noise, since you will be in the smallest "noise zone" (Figure 5).

The small, imported all-wave receivers generally use a whip antenna instead of a loop for hf and vhf reception, so the whip must be extended for this part of the search.

Due to the difference in sensitivity between a home radio or tv set and your inexpensive portable transistor radio, it is usually necessary to start your search listening to noise at a clear spot on the broadcast band dial. Tune the receiver off-station and listen to the noise level only. To speed things up, it is suggested that you start out in your car with a passenger holding the portable radio near a window. Or, you might try your car radio to start with, if you can hear the noise on it.

Start driving up and down the street, noting any increase or decrease in noise level. If you notice a drop in noise, don't immediately stop and turn around but continue on until you definitely lose the noise. Note this point, turn around (safely!) and proceed in the opposite direction. Note at which point the noise once again drops out.

Ignore Temporary Noise Increases

If you pass under a power line you might notice a sudden increase in radio noise. Ignore this, as the noise is being conducted by and radiated from the power lines. You will notice increases in noise intensity at corner power poles and transformer poles. Again, ignore these as they usually produce false noise peaks. Continue your search until the roaring noise sounds as if it will break the loud speaker. Now is the time to switch to a much higher frequency and keep searching until you can hear the interference *just below* 88 MHz on the fm band. You are now getting close to the noise source.

It is time to get out of the car and proceed on foot. With the whip antenna of the radio in a horizontal position (Figure 3), make a complete turn which should give you a line of direction along which the noise is the loudest.

Turn the volume control down so that you barely hear the noise and walk around. If you are going away from the source, the noise will quickly drop out. If it does, walk in another direction. When you feel you are close to the source, shorten the whip antenna to desensitize reception. You are closing in. Finally, you reach a point where the noise source can be in one of several buildings. It's now necessary to walk up the various driveways, sidewalks, or other access routes to determine where the noise is the loudest.

The inhabitants of these buildings would not question a teenager in the neighborhood walking up and down the streen listening intently to a radio, but they may think it a bit odd for an adult to be doing the same thing. It may be advantageous to enlist the assistance of a friendly juvenile in this task, or else be prepared to explain in a friendly way what you are doing. In the conversation, it is helpful to ask if the neighbor has experienced any radio or television interference as this may be a valuable clue.

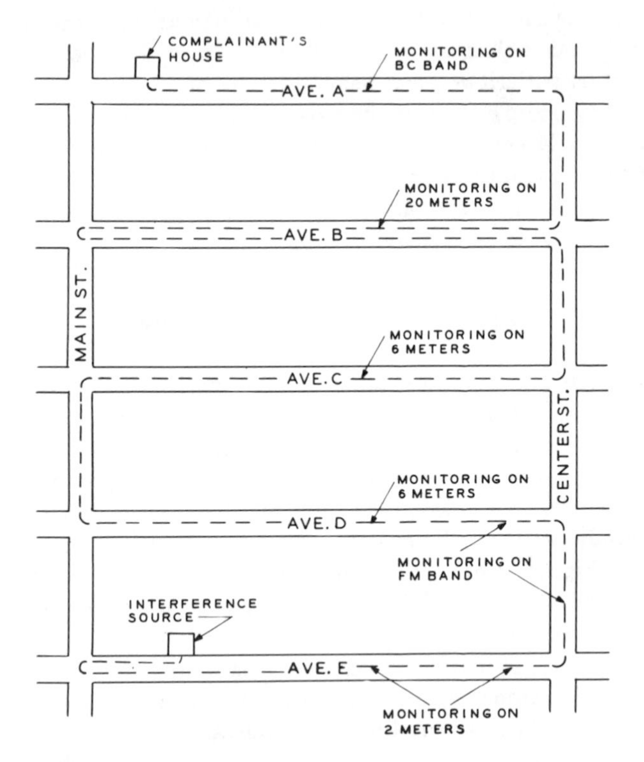

Fig. 6 Tracking noise via portable receiver and automobile. Noise level is noted as vehicle is driven along streets. Receiver is switched to higher and higher frequency as noise source becomes louder. Final search is done on foot with receiver on fm band or 2 meter amateur band.

On the Final Trail

You are now on the final trail of the RFI. Once you have pinpointed a home or building that seems to be the noise source, you should ring the doorbell, step back one step (so as not to appear intimidating to the owner or occupant) and wait for someone to appear.

Identify yourself as a neighbor, explain that you are investigating a source of radio or television interference in the neighborhood, and mention that you have narrowed the source down to several houses in the immediate area. Ask the occupant if he (or she) would be willing to cooperate with you in locating the source by making a simple test. Let the

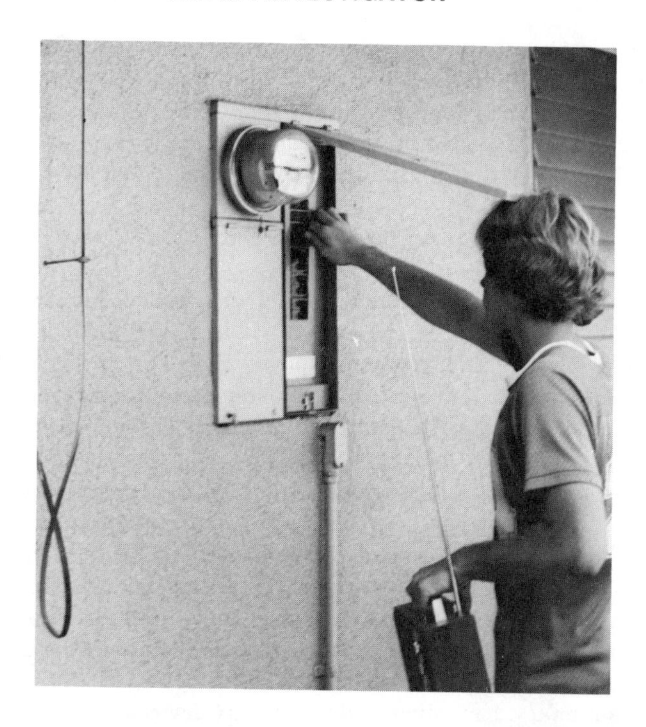

Fig. 7 With occupant's permission the electrical circuits in the residence are opened one at a time by flipping the circuit breakers or unscrewing the fuses. When the circuit serving the noisy device is opened, the interference will stop.

neighbor listen to the noise in your receiver. In some cases, they will admit to interference in their own radio or tv set.

Without going into the house, ask the occupant if he (or she) would please open the electrical circuits at his power distribution box, one at a time, by flipping the circuit breakers or unscrewing the fuses. If the circuits terminate in an outdoor box, ask the neighbor if you may follow him to the box.

If you have spotted the correct noise source, when the occupant opens the circuit that serves the interfering device, the noise will stop. (Be sure to show amazement and tell the occupant how lucky you are on this one!)

Ask the occupant to re-energize the circuit and to please disconnect everything on that circuit. Do not insist on entering the house since many occupants are sensitive about this, and all can be handled from the front door. When the offending device is located, tactfully suggest repair or replacement, thank him (or her) for cooperation--and you will probably be thanked in return.

If you happen to be a radio amateur or CB operator, now is the time to identify yourself as such. You will benefit in many ways and no longer be looked upon as "that nut down the street that ruins my tv programs!"

A Problem You May Meet

There are always some individuals who will not cooperate with you out of fear, timidity or cussedness. It is risky (and trespassing on personal property) to locate the power distribution box and pull the circuits yourself, but--with the owner's permission--it can be done. However, it is better to use the approach: "Aren't you interested in watching interference-free television?"

Most important of all, avoid an argument. Only as a last resort would you tell a recalcitrant occupant that he is operating an "incidental radiating device" that is subject to the Rules and Regulations of the FCC and that it is your duty to report him. This is an uncomfortable, abrasive, last resort and the results of this approach are unpredictable. Remember: diplomacy, politeness and tact pay big dividends in the complicated, delicate and time-consuming task of locating RFI.

In summary, remember that interfering devices are usually nearby if they are heard on a portable radio. Some may be far away if they are connected to an excellent antenna such as that furnished by fences, television antennas and power lines. Generally speaking, industrial devices and the sweep oscillator of a tv receiver can emit interference that will travel over half a mile. Neon signs, electric motors, light dimmers, fish tank thermostats and the like have been known to have an interference zone nearly two miles in diameter.Heating pads and other thermostatically controlled devices can blanket a zone of two or three city blocks.

There is always a chance you will run into a situation that does not follow these patterns; be persistent and the problem can be solved.

Wrapping It All Up

The RFI hunt can be interesting and instructive. It is not difficult to conduct, and only an inexpensive portable multiband radio is required. Most RFI sources will be nearby unless inadvertently connected to a good radiating system. In areas where the electrical service is underground, the noise zone will be cut approximately in half as direct radiation from the power line is reduced.

The Case of the Defective Light Fixture

The investigator, who was a CBer, was startled one day to hear a loud, buzzing noise on his CB set. Checking his tv receiver, he saw a small amount of interference on channel 2. A quick check indicated that the noise peaked at about 15 MHz on a portable, battery operated all-wave receiver.

Taking the portable radio in hand, the investigator walked up and down his street, finally pinpointing the noise at a neighbor's house about five doors distant. Knowing the occupants, the investigator rang the doorbell, explained the problem and let his neighbors listen to the *bzzzt-bzzzzt* noise emanating from the portable radio.

"Why, we hear something like that on our radio. And we see funny lines on the television set." Sure enough, the lines on the tv screen matched the noise pouring out of the portable radio.

The various electrical circuits in the home were disconnected one at a time at the outside switch box. The noise went off when the kitchen switch was opened and returned when it was closed. The noise source obviously was in the kitchen.

Going into the kitchen, the lights and appliances were turned off and on, one by one. But the abrasive radio noise continued. Nothing in the kitchen, it seemed, had anything to do with the noise.

All the kitchen lights were turned off. Still the noise persisted. Finally, in desperation, the portable radio was raised in the air and held near the light fixture in the center of the ceiling. A loud burst of noise indicated this was the source of the trouble, even though the lights were turned off.

The switch in the outside switch box was opened and the light fixture was unscrewed from the ceiling. Observation showed that the phenolic insulation in one socket was defective and had sparked across at some time in the past. A black, carbonized path could be seen when the socket was examined with the aid of a flashlight. Sure enough, when the socket was removed the whole underside was found to be charred and disintegrating. Since the fault occurred at the light bulb socket, the investigator was puzzled as to why the noise did not stop when the light was off. Examination of the wiring showed that the wall switch had been placed in the ground return circuit instead of the "hot" lead to the ceiling receptacle. It was decided to call an electrician to replace the defective fixture and correct the faulty wiring. No doubt the wiring had been incorrectly installed when the house was built and would have gone unnoticed except for the unexpected breakdown of the insulation in one light bulb socket.

The Case of the Defective Doorbell Transformer

The investigator was called late one night by a CB friend six blocks away who complained of loud radio noise in his CB set. Holding the telephone earpiece to the radio, the CBer let his friend listen to the intermittent, buzzing noise.

The investigator switched on his own ham rig. And there was the noise, bold and loud all over the 40 meter band. The two enthusiasts agreed to start the search for the offending noise the next morning.

It was a beautiful day as the investigator drove away from his home, monitoring the noise on his mobile ham set as well as on a portable battery receiver. He drove up and down the intervening streets between his home and the home of his friend, but the noise level remained fairly constant on the battery set. It was not until he arrived on his friend's street that he could hear the noise on the 20 meter amateur band. And by the time he pulled up in front of his friend's house, he could hear it on the 10 meter band.

The two RFI hunters got into the car, drove down the street and turned the corner. The noise grew louder and by the next block it could be heard on the portable fm receiver. Making a circle of the block, a house was finally pinpointed that seemed to be the source of the racket. Going to the front door pointed out the problem. The doorbell would not work. Knocking on the door brought the home owner, to whom the problem was explained. Acknowledging that his tv reception had been marred the past few days, the owner led the way to the switch box and distribution panel. Sure enough, there was the doorbell transformer, emitting a loud buzz. The transformer frame felt warm to the touch. One of the investigators tapped the offending transformer with the handle of the screwdriver and instantly the noise level rose and fell erratically. The transformer was at fault. Another case solved by the Fearless Investigator!

The Case of the Embarrassed CB Operator

The investigator called on a local CB operator who had complained to the power company of "bad power line interference" to both his CB set and the family television receiver. At a prearranged meeting time, the investigator met the CBer and heard the noise on the CB set. He also noticed interfering lines on the nearby television set.

When the investigator asked the CBer to turn off his set, the television picture cleared up. Further investigation revealed that a filter capacitor in the CB set had partially shorted out and was arcing intermittently inside the case. The power company had been blamed for interference created by the CBer's own set!

The Case of the Pirate Operator

The whole neighborhood was up in arms. Television reception was impossible. When the RFI investigator responded to frantic calls, the source of the interference seemed clear. "It's that darn CB operator down the street," a neighbor said. "I can hear his voice coming out of my stereo and he blanks out the television."

The investigator, after getting the facts from the excited neighbors, visited the home of the CB operator. He noted a huge beam mounted on the operator's roof and when the reluctant CBer finally let the investigator into the "radio room", the investigator spotted an illegal 500 watt amplifier that the CBer had added to his base station. The CBer protested that he never used the amplifier, reminding the investigator that he wasn't from the FCC and had no right to interfere with operation of the CB station. And furthermore, he wasn't causing any television interference among his neighbors.

The investigator was startled at this moment by the CBer's sharp-tongued wife who entered the discussion, berating her husband for wasting all his spare time "talking on CB" and also "ruining my television reception with his bootleg equipment." As the argument waxed hot, the investigator tactfully withdrew and suggested to the irate neighbors that they contact the FCC as the matter was out of his hands. They must have done that, for the investigator drove by the neighborhood a few months later and noted the giant antenna was down and the house was empty and dark. The illegal CBer had moved from the neighborhood.

Be Alert--House Wiring Can Kill!

Accidental contact with high voltage can kill or severely injure an individual. Less well known is the fact that the relatively low voltage of the house wiring can be deadly.

Single phase voltage, nominally rated at 120/240 volts, 60 Hz, is common in residential areas. It is stepped down from a distribution transformer and brought to a residence by three heavy conductors; two of

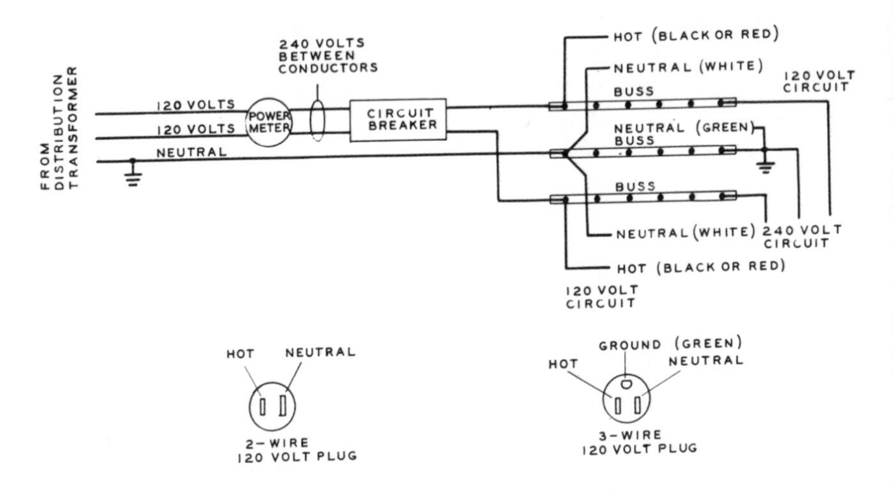

Fig. 7 Single phase circuit common to residential areas. Neutral wire is maintained at, or near, ground potential.

these wires are at high potential to ground and to each other and are called *hot wires;* the third wire is called the *neutral* and is grounded at the distribution transformer. In addition, it is grounded to the cold water pipe or a ground electrode in the residence. The neutral wire is at, or near, ground potential; the voltage across the two hot wires is 240 volts and the voltage between either hot wire and neutral is 120 volts (Figure 7).

The three power conductors are brought through the power meter and through the circuit protective devices (circuit breakers or fuses) in the distribution box to a set of buss bars. From this point, the circuits necessary to provide electricity to all parts of the residence originate. Each circuit consists of two insulated conductors, one of which is neutral and has *white* insulation; the second wire is the hot conductor and has *black* or *red* insulation. In cases where three-prong, polarized plugs and receptacles are used, a third small ground wire is used, colored *green.*

Although two and three wire wall plugs are used, as shown in the illustration for 120 volt circuits, 240 volts is also provided to power major appliances such as electric ranges, ovens, clothes dryers, air conditioners, etc. The house wiring is commonly divided between two 120 volt circuits to maintain a balance with respect to the overall 240 volt distribution system. The neutral wire provides a safety ground for larger appliances.

Every means possible is used to prevent contact with uninsulated wires and live circuits. Switches are insulated and exposed connections are covered with a face plate. Many new homes are equipped with safety plates that cover unused wall sockets to prevent children from poking metallic objects into the receptacles. In short, a formidable volume of installation and safety techniques have developed over the years and many of these features are a part of the Uniform Building Code adopted by many municipalities in the United States. Counties and cities have ordinances that specify a permit must be obtained for any work on home electric circuits. This requires that an Inspector of the Building and Safety Department inspect the wiring for workmanship and safety. Building and safety codes specify the wire size, receptacles, switches and protective devices for all residential circuits. In all cases, therefore, when rewiring or adding circuits to the home wiring, the Building and Safety Department of your municipality should be consulted in order to make sure that the wiring is in accordance with all of the rules and regulations.

In spite of all the safeguards, how is it possible for someone to get electrocuted in a residence? There are two reasons: ignorance and carelessness.

The voltage required to burn or kill a person depends upon the resistance of the contact between the conductor and the skin of the body, skin moisture and the path the electric current takes through the body. If electrical contact is made between the hands (a common case), the electric current flows through the body between the arms and directly through the heart. This can cause fibrillation (uncontrolled quivering) of the heart muscles and can be quickly fatal. The skin and tissue of the body represent a resistance value from 100 to 500 ohms, depending upon the individual. And it has been shown that a current flow of between only 10 and 250 mA can be fatal. The voltage range necessary to produce a fatal condition, therefore, can be considerably less than 100 volts when conditions are right. Line voltage, therefore, should always be regarded as *lethal.*

Aluminum House Wiring

The use of aluminum wiring in some new homes poses the threat of fire due to heat generated at the point the aluminum wire is attached to switches and other electrical fittings. While the connections were tight originally, the joint can loosen with time and corrosion may set in, forming a high resistance film over the wire. Sometimes a bad joint can cause severe RFI as well as being a fire hazard.

Safety First

How can accidental contact be prevented while doing anti-RFI work? The primary rule is that one should never work on any household circuit unless the circuit has been de-energized and cannot be inadvertently re-energized. This is accomplished by opening the circuitbreaker or unscrewing the fuse for the particular circuit. The circuit panel should be closed and padlocked to insure the circuit is not re-energized accidentally by others. Appliances, lamps, etc., must be disconnected before work is done on them.

The distribution box can also be a death trap to the uninformed. When the protective coverplate is removed, the hot busses are exposed. Because of the congested construction in most boxes, the small space around the busses and terminals is a hazard unless the box is completely de-energized by pulling the main switch. It is strongly recommended that any work on the house wiring, distribution box or power mains be done by an electrician. They know what can or cannot be done.

Caution! If you don't know what you are doing, don't do it! A rattlesnake and electricity are somewhat similar except electricity doesn't give a warning before it strikes. Treat electricity with the same respect that you would treat a rattlesnake.

Chapter 5

Power Line Interference

Sources of power line RFI and how to locate and cure them

Electric power is generated, transformed to a very high voltage for long distance transmission, transformed down to a lower voltage for local distribution, and finally stepped down to a still lower voltage for use by the consumer.

As a general rule ac (alternating current) voltage is generated and then stepped up to potentials as high as 500 kV (kilovolts) or 1,000 kV for transmission over long distances. It is then stepped down at the destination.

A different example of power transmission is the Bonneville Power Administration dam on the Columbia river between Washington and Oregon. The power is generated as ac and is then rectified to dc at 800 kV. A dc line runs to the terminal point at the Sylmar converter station in Los Angeles where the power is converted back to ac.

Electronic noise pollution can originate from the high voltage lines, from the various distribution points and from the low voltage utility lines. RFI from power lines is principally caused either by spark or corona (electrostatic) discharge. The RFI sound in a radio receiver is an undulating, frying, buzzing, scratching or popping noise. Visually, on the tv screen, it shows up as "shot lines" or "snow" in horizontal bands moving vertically up the screen (Figure 5, page 27). The width of the interference bands depends upon the proximity and intensity of the noise source.

Power line RFI can be traced to several sources: first, interference attributed to the components of a distribution system; second, interference attributed to consumer equipment connected to the power line,

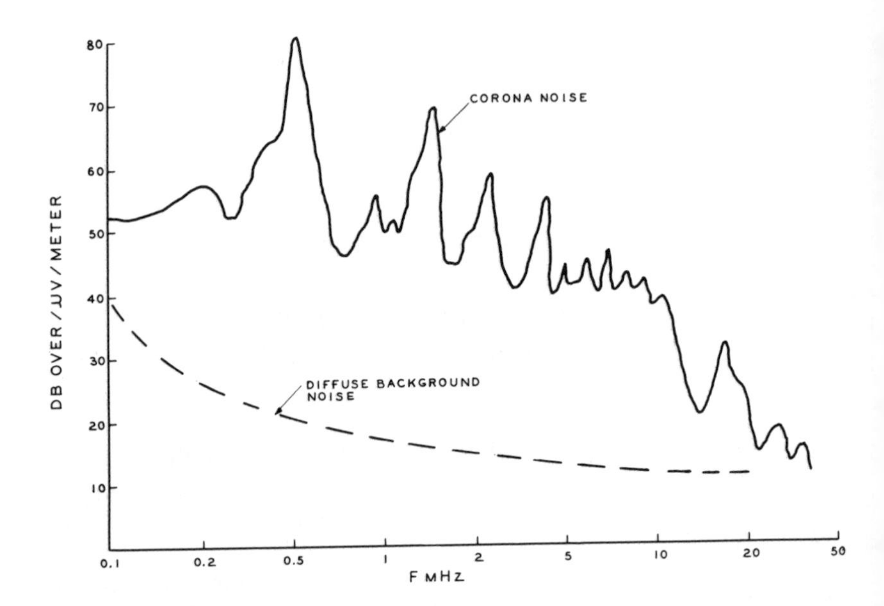

Fig. 1 Distribution of corona noise over the radio spectrum. The intensity decreases with frequency but sharp bursts of noise occur at various discrete frequencies. The amplitude of the noise increases in damp weather. The measurement shown here was made on a 230 kV ac power line in good weather.

and third, interference remotely generated and coupled into the line by normal electromagnetic propagation. The first source is discussed in this chapter.

Corona-generated RFI

High voltage lines sometimes exhibit *corona discharge* at the supporting insulators or at sharp edges or points along the wire. The discharge is due to ionization of the air in the vicinity of a high voltage conductor. Ionization is an energy transformation process, producing visible light and broadband rf energy as well as ozone. Ozone is a corrosive product which brings about the ultimate destruction of insulators and nearby metallic surfaces.

Corona streamers are velvety blue in color and start out as a glow, tightly adhering to the insulator. As ionization progresses, the glow spreads, having a paintbrush-like appearance at points on the surface of the insulator. At higher voltage stress, the corona expands and eventually changes into a spark breakdown.

Radio noise from corona discharge varies considerably with frequency (Figure 1) and with atmospheric conditions, being greater in intensity in damp weather. Bursts of noise occur at discrete frequencies, with the overall noise amplitude gradually decreasing with increasing frequency.

Corona noise can be reduced by proper attention to power line construction and the elimination of dirt and contaminants on line insulators. Many of the remedies mentioned in this chapter for elimination of spark interference apply equally well to corona interference.

Power Line Sparking

A common source of noise is sparking at some point along the power line. A spark occurs at sea level atmospheric pressure when there is an air gap of less than .05 cm between metal components or hardware on a power pole. The metal surfaces need not be connected to the lines since induced voltage from the lines can charge the nearby metals to a sparking potential of about 300 volts. The spark is composed of broadband rf energy and for a 60 Hz line has a repetition rate of 120 Hz. The spark from hardware on an overhead power line causes RFI over a large area as the lines pick up the spark energy and propagate it for a great distance.

Generally speaking, most sparking RFI generated in an overhead power system originates on wood poles which carry distribution voltages from 2.4 kV through 55 kV and transmission voltages from 60 kV through 115 kV. In most cases, the higher the line voltage, the less the noise. Voltages above 138 kV tend to be less noisy than lower voltages because these circuits are carried on steel structures and are designed to eliminate or reduce noise sources when the lines are built and energized.

The Power Line Spark

A power line spark is created by the process shown in Figure 2A. A spark cannot take place unless there is a high resistance (R1) between the power source and the spark gap. This resistance can be an oxide of metal (rust, for example) on the pole hardware, the electrical resistance of the wood of the pole, or the leakage path across an insulator--or a combination of all of these.

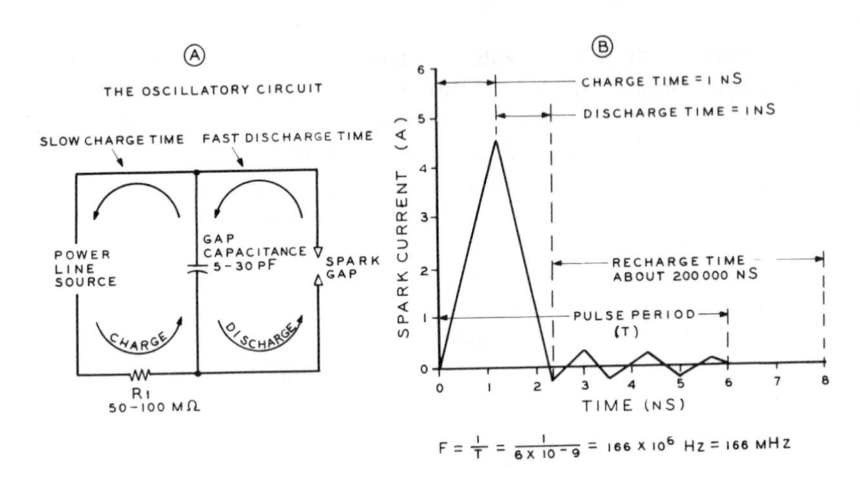

Fig. 2 The corona oscillatory circuit (A). When the gap potential is great enough a relaxation oscillator (RC) circuit is formed. The period of the pulse train is about 6 nanoseconds and the equivalent frequency is 166 MHz. Repetition rate is 120 Hz for a 60 Hz line.

When the potential difference across the gap is great enough and the series resistance is high, a relaxation (RC) oscillator circuit is formed in which current alternately flows and ceases to flow across the gap. A molecule of air in the gap will be ionized when the gap potential is sufficient to impart enough kinetic energy to a free electron to cause it to knock other electrons from their orbits within the atoms of the air molecule. This results in freeing more electrons and creating positive ions. The collision process continues until the gap is sufficiently ionized to cause a rapid decrease in gap resistance from thousands of ohms to about 200 ohms. When this occurs, an avalanche of electrons flows across the gap. Also released when electrons are knocked from their orbits is a photon of light. Oxygen's photon is blue and accounts for the bluish color of the spark.

A tremendous supply of electrons is obtained in a very short period as shown in Figure 2B, the spark current building up to a maximum value in about 1 nanosecond (10^{-9} second). The discharge time of the gap capacitance is about the same period. The short rise and fall times of the spark indicate a release of energy into the radio spectrum. If the example discussed is considered, the period of the pulse is about 6 ns and the equivalent frequency of the discharge is approximately 166 MHz. The intensity of the RFI depends upon the size and character of the gap, the voltage breakdown point, the moisture content of the atmosphere, and the radiating efficiency of the "antenna system" composed of the power lines.

The preponderance of RFI caused by power systems is centered around wood poles, crossarms, and the metal hardware used to secure the various components to the pole--and by poor construction practices. Contrary to popular thought, little RFI is caused by steel poles that carry very high voltage lines. In most parts of this country, voltages carried on wood poles range from 120 kV downwards, while higher voltages are carried on the steel poles.

Poor Construction Practices

Many of the poor power line construction practices of the past which create the RFI problems of today were the result of inadequate training of construction crews, particularly during the period 1955 to 1970.

As far as sparking was concerned, the maintenance crews were only instructed to climb the suspected pole, tighten the hardware, bonding staples and tie wires and to work over the dead ends. They had no way of telling whether their efforts produced the desired results. Power companies, too, were not alert to the RFI problems created by loose hardware and often turned a deaf ear to complaints from residents living near the lines whose radio reception was jammed by heavy line interference.

The learning process was long and painful. Many sources of power line RFI have been discovered that were not obvious to the untrained construction crews of yesterday with no knowledge of RFI.

In the past did anyone tell the lineman that a loose tie wire would cause RFI?

*Did anyone tell him that an insulated tie wire on a bare conductor would cause RFI?

*Did anyone tell him that a bare tie wire tag left on a transformer hanger and conductor would cause RFI?

*Did anyone tell him that the retaining ring that slid off the automatic dead-end would cause RFI?

*Did anyone tell him that a square washer left on the space bolt between the double arms would cause RFI?

*Did anyone tell him that span guys contacting each other or touching communication cables could cause RFI?

*Did anyone tell him that a bond wire crossed over itself between the double washers could be broken and would cause RFI?

*Did anyone tell him that excess bond wire wrapped around the space bolt could cause RFI? Or that several staples bunched together over the bond wire could cause RFI?

*Did anyone tell him that a bond wire installed without double washers would cause RFI? Or that the nut holding the bond wire between the washers must be more than finger-tight or it would cause RFI?

Much of this information was learned the hard way over a period of years, but apparently little of it was imparted to the construction and maintenance crews. If time and effort had been spent instructing construction crews in the sources and causes of RFI, many power companies would not be confronted today with the costly RFI problems which continually arise.

The Wood Power Pole

Silhouetted against the sky but little noticed by the citizenry over the years, the common wood power pole and crossarms are the mainstay of the suburban electrical system. As the pole ages, interesting changes take place which eventually lead to serious RFI problems.

From season to season the wood expandsduring the rainy periods and contracts during dry, warm days. The bolts, nuts and metal hardware also expand and contract to a lesser degree. Vibration and shrinking of the wood gradually loosen the bond staples holding the bond wire to the crossarm or pole. Nuts and washers loosen. The pole settles in the earth and the conductors become slack and sway in the wind. The oscillations of the conductors wear out the tie wires on the insulators and cause vibrations to be set up in the crossarms which may not be in sympathy with other vibrations of the pole. And all the while oxides (rust) form on the metal hardware.

Fig. 3 Closeup of pole hardware. 1- Loose bond which can result in a minute arc between bond wire and staple. 2 and 3- Square washer with spring washer to maintain tension. 4- Bond wire crossed over itself could break between the washers and create RFI.

Creation of the Rf Spark

The metal hardware on a wood pole is situated in a powerful, alternating electric field set up by the line conductors. It may also be in a strong magnetic field due to the load current flowing in the lines. Distortion of the fields by the metal hardware causes voltage stress to be

Fig. 4 Tightening loose hardware is a quick (but impermanent) cure for an RFI problem when only one crossarm is on the pole, but it becomes a time-consuming, expensive task when multiple crossarms are involved.

set up in regions of close proximity (Figure 3). Breakdown and spark discharge occur across small air gaps or through the corrosion film between two pieces of hardware. In addition to coupling the resulting interference back into the power line, these pieces of hardware can form an efficient dipole antenna at certain frequencies, radiating the interference to the area about the power line.

In a typical case, the magnetic field about the line conductors induces voltage into the metal hardware in the field. A washer may become loose after a period of time and current flows between it and the nut and bolt. The current flow is hampered by the resistance of the surface oxides on the hardware. Slowly, voltage builds up between the washer and the nut until a point is reached where there is a discharge through the oxide, creating an inaudible spark, rich in wideband rf energy. This minute rf voltage is fed into a nearby "antenna system" composed of the overhead conductors and is radiated or conducted into audio and video devices in

Fig. 5 New armless construction, esthetically pleasing, uses pole-mounted brackets and insulators. Unless spring washers are used under the nuts the bolts can work loose and cause severe RFI.

the area. "Shot lines" appear on television screens and a frying, buzzing noise knocks out radio reception. On some days the RFI is absent, on others it is very bad. When it is on, neighborhood radio amateurs and CBers are often blamed for the interference.

Finally, after tempers are frazzled, an exasperated neighbor calls the local power company and an investigator goes out to look into the RFI problem. When he finally locates the source of the racket on a particular pole, he reports it and a line crew is scheduled. A lineman climbs the pole, tightens the hardware, hammers the staples back into the wood, tightens the tie wires and departs. The source of the RFI has been eliminated--for the time being.

Tightening loose pole hardware is a quick (but impermanent) cure for an RFI problem when only one crossarm is on the pole, but it becomes a time-consuming, expensive task when multiple arms are involved (Figure 4).

Fig. 6 Circulating current in loose hardware can char wood under the fixture. Hair pin washer is in use on millions of poles as a replacement for the old spring washer. When hair pin washer cracks (as shown here) it results in an arc which causes more RFI problems and scorched poles.

Modern Power Poles

In recent years efforts have been made to make power lines and poles more pleasing to the eye. Crossarms have been replaced by an armless type of construction making use of pole-mounted brackets (Figure 5). The bracket and insulator are attached to the wood pole with sturdy bolts. However, in many cases, the weight of the conductors and movement of the pole and wires gradually forces the base of the bracket into the pole. As a result, the washer becomes loose and works against the bolt head. The intermittent contact is a prolific source of RFI as the surfaces gradually oxidize and form a miniature spark gap.

Burning Poles

In some cases, circulating current in loose hardware can cause burning or charring of the wood under the hardware. This damage often

Fig. 7 New double helix spring washer reduces the problem of loose hardware and RFI and is currently being used on most new power pole installations.

occurs on a pole supporting three-phase transformers, capacitor banks and associated multicircuits. The magnetic field about these is coupled into the adjacent metal hardware with unfortunate results. To prevent burning, power engineers realized that all hardware must remain tight. This was difficult to accomplish over a period of time because of normal wood shrinkage and vibration. Lock washers would not do the job so a hair pin spring washer was eventually developed to replace the common square washer (Figure 6). The new washer was substituted for the old in the hope that the spring tension would hold the hardware secure. The hair pin washer is used today on millions of wood poles in this country. Unfortunately, time has proven that the hair pin washer eventually works loose or splits at the bend of the hair pin, creating more RFI problems and scorched poles.

Finally, an economical means of keeping hardware tight was found: a double helix spring washer (figure 7). This device, when installed between a nut and a flat washer, reduces the problem of loose hardware and RFI to a marked degree and is currently being used on most new power pole installations.

The Plastic Washer

The rounded surface of a pole is cut and chisled out to provide a flat surface for the crossarm. (This surface is called a *gain* in the industry). A

Fig. 8 Plastic "gain" is used as an insulator between crossarm and pole. It is made of high impact plastic and prevents crossarm burning as well as reducing resulting RFI.

plastic gain has been developed that provides a flat surface for the crossarm. The device is made of high impact, high dielectric Delryn plastic and forms a tough cushion between crossarm and pole. The plastic gain has been used by several utilities throughout the country which have found it has prevented crossarm burning as well as reducing resulting RFI. A typical gain is shown in Figure 8.

The next step was to develop a plastic washer to be placed under a bolt head. A section of power line was rebuilt using a new, opaque plastic washer and periodic RFI checks were made over a period of five years (Figs. 9 & 10). No RFI was noted and the installation remained mechanically tight. Finally several of the plastic washers were removed for examination. Except for a slight deformation, they were in satisfactory condition.

The final washer design makes use of a black Delryn plastic to reduce deterioration of the material from ultraviolet light from the sun. The new washers and gains are a big step toward the eventual elimination of powerline RFI caused by loose hardware discharge and it is hoped these devices will be in universal use in the near future.

Fig. 9 These improved plastic washers are used under a bolt head on new construction. The washer helps eliminate powerline RFI caused by loose hardware discharge. Opaque washer reduces deterioration caused by ultraviolet light from the sun.

Powerline Insulators

An *insulator* is a device which resists current flow between conductors. Most power poles use insulators made of porcelain. The insulators are designed so that the "creep distance" or electrical path along the surface is great enough to prevent flashover (Figure 10). The skirts and flutes of the insulator make it possible to have a long creep distance on a small insulator. The voltage holdoff required by the circuit determines the size and shape of the insulator.

A different style of insulator is used for distribution voltages as high as 33 kV. The *bell insulator* (Figure 11). is commonly used for deadending conductors to a crossarm, as shown in Figure 12. The line voltage determines the number and size of insulators required.

A bell insulator is a very strong device but can be a source of prolific RFI. When the tension on an insulator string is lessened because of slack wires, the clean metal-to-metal contact between insulator sections becomes faulty, oxides are built up and tiny spark discharges occur across the insulator hardware. It is difficult to maintain a taut span of wire over the years as wire stretches and poles settle into the ground. In addition, the power pole slowly leans in the direction of greatest wire

Fig. 10 (top) Post-type insulator resists current flow between conductors. Skirts and flutes of insulator provide a long electrical path. Fig. 11 (botton) The bell insulator is used under tension to attach a conductor to a crossarm. Metal end connections allow two or more insulators to be connected in series for long "creep path."

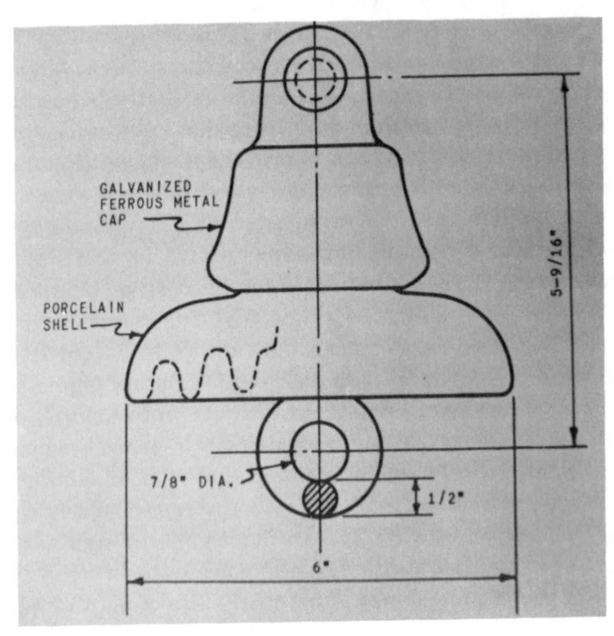

Fig. 12 Seriesed bell insulators are used to deadend conductors to a crossarm. Three wires are terminated in this fashion. Notice that the wires are slack and sparking can occur across insulator-connecting hardware due to induced voltages.

tension. The result of these movements is that the bell insulators become slack and RFI from induced spark discharge is noticed soon after. Repulling the conductors to provide proper tension will not cure the RFI once it has started and it is not economically feasible to replace each set of RFI-generating hardware because sooner or later the problem would recur. Thus it became evident some years ago that a need existed for a corrective technique that would stop the RFI, regardless of the amount of slack in the insulators.

Reworking the Bell Insulator

An inexpensive solution to the RFI problem associated with the bell insulator is the application of conductive grease (containing particles of graphite) to the insulator hardware. This completely knocks out induced

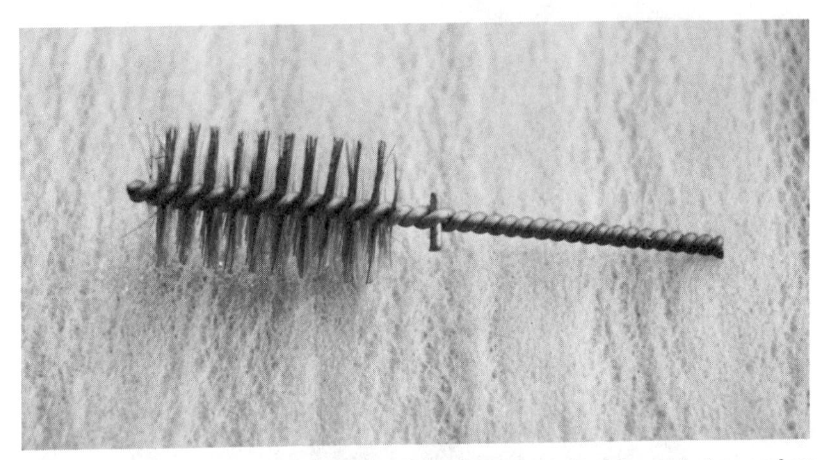

Fig. 13 "Rifle Barrel" bonding brush is inserted in hardware between two bell insulators to provide a path for flow of induced current. The brush bristles are made of stainless steel and make good electrical contact between metal portions of the insulators.

RFI for a period of time until the grease hardens or is washed away. A long-term solution to the problem was the creation of the *bonding brush*, which resembles a rifle barrel cleaner (Figure 13). The brush is about 3/4-inch in diameter and made of fine stainless steel wires. The brush is inserted in the pin and clevis between two insulators and provides a semi-permanent means of controlling the flow of induced current. Under normal conditions, the brush is good for about three years before it flattens and falls out, or is washed out during insulator washing by maintenance crews.

The RIV Clip

The next step in the battle against RFI created by slack insulators is the *RIV* (radio influence voltage) clip. This device is made of stainless steel (Figure 14) and is easy to install in the area between two bell insulators. When used in conjunction with bonding brushes, the combination provides RFI-free service for up to five years, after which time the clip loses tension and must be replaced at considerable expense and inconvenience.

An improved RIV clip is shown in Figure 15. This clip is made of specially treated phosphor-bronze material and is formed in such a way that it locks around the hardware in the bell insulator. Most clips in use today, however, are the stainless steel type.

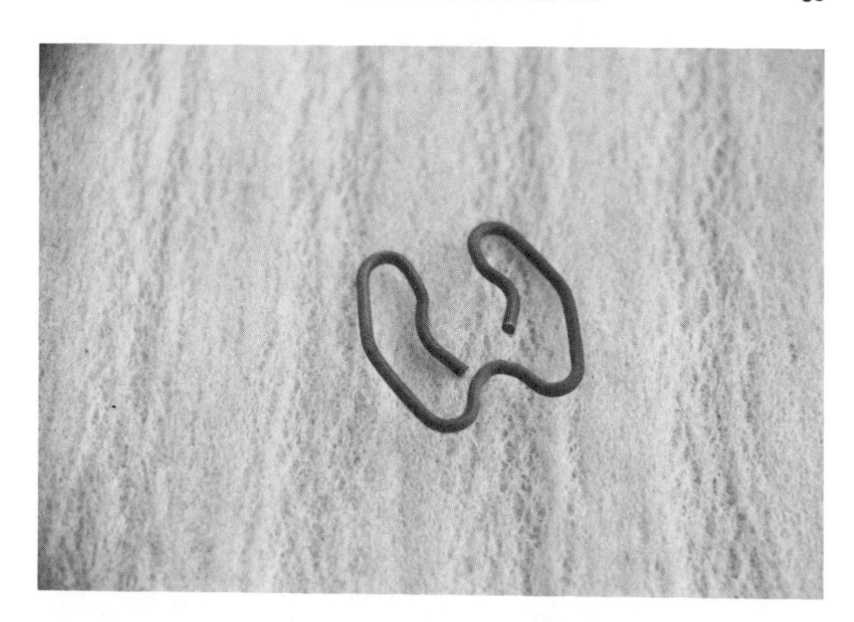

Fig. 14 (top) RIV clip is made of stainless steel and locks around the hardware in a bell insulator. Fig. 15 (bottom) Improved stainless steel RIV clip corrects and eliminates RFI when properly installed.

Fig. 16 One-piece post insulator has no separate metal parts and eliminates sparking common to slack bell insulator.

New Insulator Designs

A recent replacement for the bell-type deadend insulator string shown in Figure 12 is a post-type insulator mounted in the horizontal position (Figure 16). This one-piece unit has a metal base and metal cap onto which is fastened the deadend shoe. Installation of the horizontal post insulator eliminates three potential sources of RFI from the two-unit bell deadends and five potential sources from the three-unit deadends. Since there are less metal parts in the post-type insulator the gradual reduction of line tension does not cause problems and the use of this post insulator reduces RFI problems associated with slack deadends.

Not all bell-type deadends should be replaced with horizontal post-type deadends. The proper use of the bell and horizontal post insulators is shown in Figure 17. In this case there is a wire change size at the pole. The large conductor on the right meets the smaller one on the left and it was necessary to deadend on both sides of the crossarms with span guys to back up the larger sized conductor to equalize the difference in tension.

Fig. 17 Seriesed bell insulators are used on conductor at left and post insulators are used on conductors at right. One-piece post insulator eliminates RFI problems caused by loose insulator hardware if it is installed properly.

Fig. 18 A flexible strap welded between the metal surfaces of bell insulators provides a permanent electrical path. Insulator at left also has strap welded to metal washer used in installation of other hardware.

Fig. 19 Crossarms ready for installation show bonding wire running along the arm, connecting all related metal parts. Staple gun sits atop center crossarm. Note that insulated staples are used.

Bonding Insulators

A different approach to the problem of sparking hardware on power poles is to bond all metal surfaces together. A flexible strap welded between the metal surfaces of bell insulators reduces the RFI problem as it provides a permanent electrical path between insulator surfaces. Bonded insulators, such as shown in Figure 18, are commonly used in modern power line construction, but the cost of bonding existing installations is prohibitive.

Crossarm and Pole Bonds

Insulator contamination from dust, dirt, salt spray and rust is a serious problem and an ever-present source of RFI. The leakage path from conductor to crossarm and from crossarm to hardware can cause burning due to the heating effect of the current flowing through the wood, or because of resistance loss in the hardware. Sometimes the cumulative

Fig. 20 Insulated staple used to attach ground wire to pole. The plastic material prevents the staple from touching the wire.

effect is enough to cause burning which destroys the crossarms and the top of the power pole.

To prevent burning, all metal parts related to the conductors (even though the metal is insulated from the conductors) are bonded together on poles carrying voltages of 12 kV and higher. The bond wire connecting the metal parts prevents the flow of current on the surface of the crossarm and pole from reaching locations where high resistance occurs, such as the thermal resistance between the through bolt and its hole in the pole, or the dry points between the crossarm and pole at the point they are joined together.

A common bonding technique is to run a No. 8 bare copper wire (or a No. 6 bare aluminum wire) along the crossarm, stapled to the underside of the arm, connecting all related metal parts. A heavy-duty staple driven by an air gun is commonly used (Figure 19).

Unfortunately, over a period of time the bonding staples work loose from the wood, nullifying the effect of the bonding wire as far as RFI is concerned. Oxides form between the bonding wire and the loose staple and a minute spark discharge takes place between the two. A temporary

Fig. 21 Section of bonding wire showing how plastic insulators prevent contact between staple and wire. Staples are rapidly fixed in place by using an air-driven staple gun.

cure is for a lineman to climb the pole and pound the staples back into the wood so that they once again made contact with the bonding wire.

The ultimate solution to this vexing problem is to install a bonding wire with insulated staples(Figure 21). Insulated staples are now coming into general use and have resulted in a bonding system which provides a significant reduction in powerline RFI. The insulated staples are also used on lightning rod installations which require a grounding wire running the length of the pole.

Insulator Contamination

The effect of surface contamination was mentioned briefly in connection with crossarms and hardware. The gradual buildup of dust, smog and salt spray on power line insulators eventually provides a conductive path over the surface of the insulator that produces RFI. This form of noise is most noticable when there is moisture in the air and usually decreases when the sun comes out and dries the insulator.

·A temporary fix is to hot-wash the insulators periodically with a.

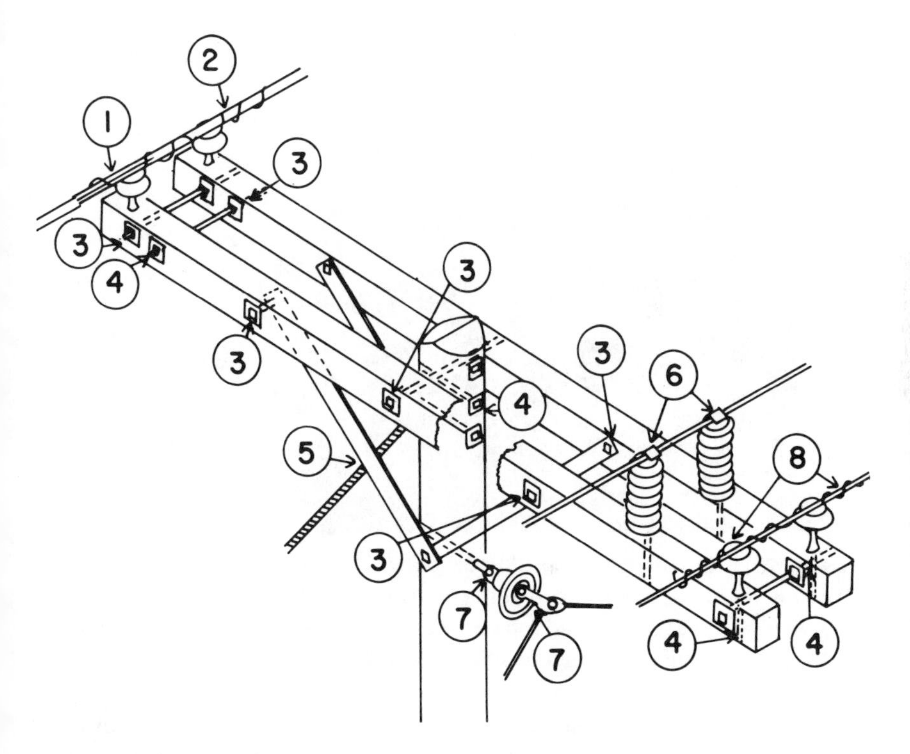

Fig. 22 Some of the possible sources of RFI on a power pole. 1-2: Weatherproof or insulated tie wires used on bare conductors or bare tie wires on insulated conductors. 3- Points of loose hardware. Sparking occurs between nut and washer. 4- Sparking between nonassociated hardware. One and one-half inch clearance is required. 5- Guy wire sparking to adjacent crossarm brace. 6- Loose clamps on top of insulators. 7- Loss of clean metal-to-metal contact in pin and clevis area of bell insulators. 8- Loose tie wires.

high pressure stream of water. Circuits along highways can be hot washed, but those running along the rear property lines cannot be reached by a water truck and must be cleaned by hand, a costly process and done only in severe cases of RFI.

Silicone grease is often applied to power line insulators in an attempt to reduce leakage, flashovers and resulting RFI. This grease encapsulates the existing particulates contaminating the surface of the insulator and provides a surface where water cannot form a film, especially during a heavy fog.

Fig. 23 Tie wire holds conductor to insulator in typical installation. The arrow points to a gap between the tie wire and the conductor. The gap can create RFI caused by sparking between the turn and the conductor. Proper wrapping reduces chance of RFI developing at this point.

Silicone grease appears to provide insulator protection for about five years after which a new coating must be applied. A method has been worked out to spray silicone grease on energized lines which seems to be very effective in reducing RFI problems.

Tie Wires

The purpose of a tie wire is to hold a conductor to an insulator. The lineman lays the tie wire across the conductor and with a twist he wraps it around the neck of the insulator in opposite directions. He then wraps the free ends of the tie wire around the conductor. And he should know he must use weatherproof ties on weatherproof conductors and bare ties on bare conductors. Typical tie wires are shown in Figure 22, but note that weatherproof ties are used on bare conductors--not a good practice!

Wrapping a tie wire properly reduces the chances of RFI developing at this point at a later date. The lineman may assume that if the tie holds the conductor to the insulator the job is done properly. This is not true because if one turn of the tie wire is loose enough to leave a gap between the wire and the conductor it could create RFI caused by a minute spark

Fig. 24 An intermittent contact between switch arm and a galvanized extension rod can cause RFI. Bonding metal parts helps reduce noise. A newer switch design having a fiberglass extension rod eliminates the RFI problem from this source.

between the turn of the tie and the conductor. To eliminate the RFI, someone will have to climb the pole at a later time and retie the conductor.

Tie wires are made of soft-drawn aluminum or copper wire and in most cases, over a period of time, even the tightest tie wire will work loose. When the tie wire is worn by vibration and oscillation, oxides form between the wire and the conductor and a microdischarge takes place.

In recent years a preformed, vibration-proof, semiconducting tie wire with a neoprene jacket has been developed. The portion of the tie wire in contact with the insulator is coated with neoprene but the ends of the wrap are left bare. This device has been under trial for several years and is now coming into general use across the country. Use of the new tie wire has resulted in a marked decrease in RFI problems in the vicinty of the power line.

Pole Switches

Gang-operated switches can be seen on various poles for the purpose of controlling power circuits. The switch mechanism commonly consists of a rocker arm and extension rod that permits manipulation of the switch from near ground level. Not much RFI is associated with these switches unless the power line parallels a high voltage transmission circuit; if it does, it is necessary to bond across the rocker arm to eliminate voltage-induced sparking and resulting radio and tv interference. Some installations eliminate the metal arm and rod and substitute a fiberglass rod passing through pole-mounted guides (Figure 24). This new design completely eliminates the RFI problem from this component.

New Hardware for Power Poles

Spark discharge between metal hardware components on power poles is a serious, continuing problem. Coupled with power leakage caused by salt spray or other contaminants, the charring and burning of poles is an expensive burden for the power company to bear. The ultimate solution is to insulate all metal parts on the pole, or bond them together. Bonding techniques have been developed to do this task. Continuing research into pole installation and the reduction of induced voltage in hardware will, in the long run, spell the end of power pole RFI. In the meantime, hundreds of thousands of old power poles are scattered across the landscape. These old installations will continue to be a source of RFI for years to come.

Aluminum Bus in Substations

As the price of copper has soared in recent years, aluminum has been used as a substitute in home wiring and in power stations, such as shown in Figure 25. This is typical of many installations.

The aluminum power bus is fastened to the supporting framework by means of an insulator and a loose-fitting aluminum alloy, doughnut-shaped clamp that permits the bus to shift about with changes in temperature and load (Figure 26). The aluminum bus is economic but can be a

Fig. 25 Aluminum conductors have been used as substitute for the more expensive copper in many substations. Aluminum oxides make poor conductors and result in junctions that create severe RFI.

prolific source of RFI to nearby communication devices because of oxides formed between the bus and the supporting clamp. Installations such as these are in constant use and will remain so for many years. The cost of modifying the supports is heavy and the chance of rework to reduce RFI is remote.

Training Power Line Crews

It is a gratifying experience for the lineman on the pole to hear that his efforts to correct a source of RFI have been successful. This confirmation is seldom possible because the RFI investigator cannot always be present when a problem is being corrected. Ideally, time should be allotted for a group of linemen to travel with an RFI investigator to a pole with a known source of RFI. Letting the linemen hear the RFI and observe the results of various methods used to suppress the noise is money well spent.

It may be a good idea to include a radio receiver in service and patrol trucks in order that linemen can directly check the results of their RFI-correction work. This is a good idea if the source of RFI can be depended

Fig. 26 Loose-fitting aluminum clamp permits bus to move about with changes in temperature. RIV clip installed between bus and clamp maintains solid contact between the two components.

upon to be active when the crew arrives! As often as not, the source will arbitrarily stop of its own accord just on the day the crew arrives to do corrective maintenance, thus they may receive the impression they are working on the wrong pole.

In any event, progress has been made over the last few years in developing economical methods of correcting and preventing RFI caused by overhead power facilities. The electric utility should have a genuine interest in good customer relations and do everything within the limits of reason, sound economy and good engineering practice to eliminate or reduce the level of RFI created by its facilities. Progress is good in this direction. Remember that because of the complexity of power line RFI it can often take considerable time to correct the problem.

Sources of RFI--reducing Hardware For Power Lines

RIV Clip (TVI-SLC-6) and RIV Clip Applicator (TVI-CIT) are available from: Utility Sales and Engineering Co., P. O. Box 6407, Orange, CA 92667. (714) 970-1101.

Insulated staples and plastic insulators are available from Utility Sales and Engineering Co. (see above).

Plastic washers are available from: Radar Engineering Co., 9535 N.E. Colfax Ave., Portland, OR 97220. (503) 256-3417.

Glossary of Power Line Terms

Automatic deadend: A device used to hold a conductor to a deadend. The wire is inserted in a tube where it is held by internal grips.

Bonding wire: A wire stapled to the crossarm or pole which connects all related hardware together, such as those devices that support the conductor.

Clamp top: A mechanical device used in place of a tie wire to hold a conductor to an insulator.

Crossarm: A straight grained rectangular wood support attached in a horizontal position on a wood power pole.

Capacitor bank: Several large fixed capacitors installed on a circuit to improve the power factor.

Cutout: A device similar to a fuse holder and fuse that is used on high voltage circuits.

Deadends: Insulators used to terminate a conductor to a crossarm or pole.

Down guy: A stranded steel wire that runs from the top of a pole to an anchor in the ground. It is sometimes called a deadman.

Eye bolt: A bolt with an eye at one end to which deadends are attached.

Hogeye: The same as an eyebolt except that it is secured to an existing bolt on the crossarm or pole.

Lightning arrester: A device used to divert lightning to ground when a strike hits a circuit.

Pothead: A means to terminate underground conductors to overhead conductors.

Span guy: A stranded steel wire that runs from pole to pole. It is used to hold tension when there is a change in conductor size.

Tie Wire: A short piece of copper or aluminum wire used to secure a conductor to an insulator.

Synchronous Switching Devices

In addition to line generated noise a second noise source exists in urban areas that is a great cause of interference, especially to broadcast reception and low frequency signalling and location devices. This annoying, wideband noise has been attributed to "power line harmonics" for many years but examination has shown that this racket does not originate from standard utility system components but from customer-related sources which feed impulse energy back into the utility system. The power lines carry these impulses along with the electrical energy.

The noise impulses are generated by high power synchronous switching devices (triac, diac, SCR diodes, etc.) that are employed in industrial process controls. The noise pulses are synchronized to the power line frequency with harmonic components spaced at 60 and 120 Hz for single phase circuits and 180 and 360 Hz for three phase circuits.

Literally thousands of these noise sources are located in a large city and unless power line filters are used on these devices, the switching transients pass freely into the power system. The great majority of such devices that have been examined have no such filters.

This is a nationwide problem and no simple solution exists at this time. It is mentioned to emphasize that not all power line noise is caused by utility lines and components but rather by devices attached to these lines. RFI generated by these synchronous switches is beyond the control of the local utility.

Manufacturers of RFI/EMI Filters

Amp Inc., Box 3608, Harrisburg, PA 17015
Axel Electronics, 134-20 Jamaica Ave., Jamaica, NY 11418
Bell Industries (J.W. Miller div.), 19070 Reyes Ave., Compton, CA 90221
Cornell-Dubilier Electronics Corp., 150 Ave. L, Newark, NJ 07105

Erie Technological Products, Box 961, Erie, PA 16512
Genisco Technology Corp., 18435 Susana Rd., Compton, CA 90221
ITT Cannon Electric, 666 E. Dyer Rd., Santa Ana, CA 92702
RFI Corp., 100 E. Pine Air Dr., Bayshore, NY 11706
Sanders Associates, Inc., Daniel Webster Highway, Nashua, NH 03060
Sprague Products Co., 481 Marshall St., North Adams, MA 01247
Tobe Deutschmann Labs., Box 74, Canton, MA 02021
TRW Inductive Products, 150 Varick St., New York, NY 10013
Watkins-Johnson Co., 3333 Hillview Ave., Palo Alto, CA 94304

A Hunt for Power Line Spark Discharge RFI

One of the most common types of RFI is the spark discharge which shows up on a tv screen as shot lines and in a receiver as a buzzing, frying sound. Because of the characteristic sound, this type of RFI is easily diagnosed as not coming from a radio transmitter.

An interesting example of an unusual RFI hunt was started by a letter received by a power company a few years ago complaining of severe radio interference. The letter said, in part, "About a week ago there was a power blackout in my area which lasted for about two hours. When the power came on again, I could not receive any broadcast stations because of the static. This is the baseball season and the team I'm rooting for is in the middle of the pennant race. Please help me out!"

It was the investigator's task to locate the mysterious RFI. The complainant lived in the rolling hill countryside far from obvious sources of RFI. He had all the modern conveniences and the area was planted in citrus and grapes: an ideal spot, far from the lifestyle and noise of the big city.

The district personnel of the utility company had been questioned regarding the power blackout and the investigator was told that two crews had patrolled the area and no cause for the breakdown, or subsequent noise, had been found. The investigator believed that if he could trace the RFI he could also locate the cause of the circuit outage. He requested that a maintenance crew accompany him on the search (Figure 27).

Driving into the vicinity, with the car radio tuned to a clear spot near 640 kHz, the investigator heard the noise about a mile from the ranch house. He drove past the home and explored a branch power line feeding from the main line, but the noise soon began to drop out. At the next branch (or lateral), he began to hear the RFI on 80 meters. He drove along

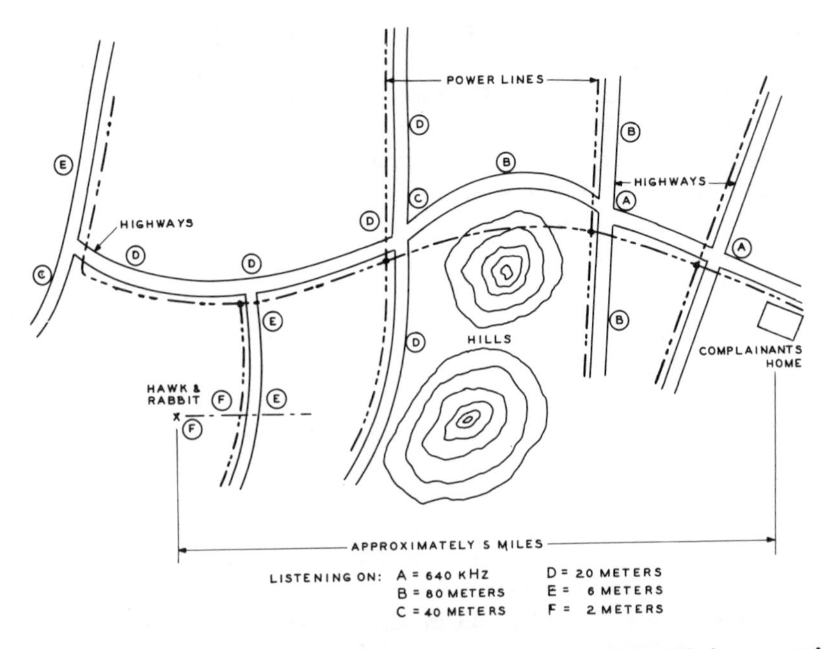

Fig. 27 Noise hunt starts on broadcast band at complainant's house and progressively advances through hf ham bands until RFI is noticed on 6 meters at point E. Interference source at F was found on 2 meters.

the line, hearing the noise as high as 80 meters, then turned back and drove in the opposite direction, noting that he could hear the noise on 40 meters at one spot between two hills. Switching back to 80 meters he followed the main line until it went over the hill. When the line came down from the hill and again followed the road, the investigator began to hear the noise on 20 meters. He drove to the end of the line, listening on 20 meters, and as he passed one of the branches he noted a much higher noise reading on the S-meter of the receiver. The investigator then drove along this suspicious lateral and soon was able to hear the noise on the 6 meter band. Finally, he picked up the noise on the 2 meter band; the RFI peaked sharply at a pole supporting three transformers which served a wind machine used to protect the citrus trees on cold nights.

The investigator got out of his car and walked over to the pole. At the foot were a large, slightly singed, dead hawk and a small dead rabbit. The ill-fated hawk held the rabbit in its claws. The bird apparently grew tired in flight and landed atop one of the transformers for a rest. In landing, the hawk's wings must have touched the primary cables exposed at the transformer terminals. There was a short circuit through the body of the

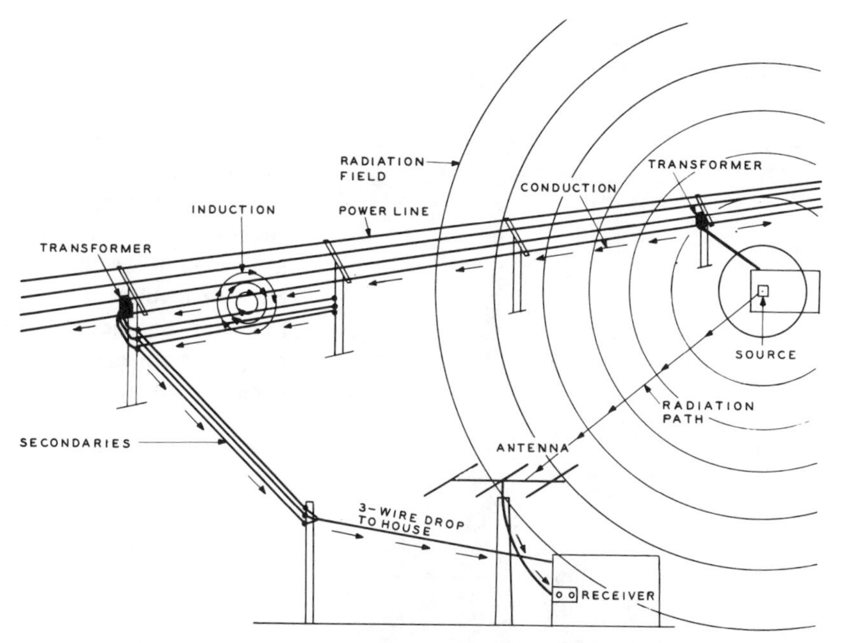

Fig. 28 Powerline RFI is conducted along the lines but may be radiated directly from the source as well as by the lines. Inductive noise coupling between lines can exist as well. Identification of the source is a time-consuming task.

bird, killing it instantly and tripping a sensitive relay at the substation which opened the primary circuit breakers.

The flashover when the hawk hit the wires damaged one of the bushings on the transformer and an audible spark could be heard from the ground. Two of the three protective fuses had blown, thus leaving one of the phases energized; the RFI was created by the spark discharge between the bushing and the transformer case. When the other fuse was removed by a maintenance man, the RFI stopped.

A hole was dug and the remains of the two victims were buried and everyone hoped that the souls of the animals had entered their particular Valhalla.

As the hawk flies, the RFI source was 5 miles from the complainant, but over 15 miles of driving were required to locate the source. The RFI had been radiated from the source into the power lines and then re-rediated to the rancher's radio receiver.

Fig. 29 Expert can identify individual power-lines on this large pole. Highest voltage lines are at the top. Note different style of insulators used. RFI may be inductively coupled from one set of lines to another.

Powerline RFI- A Summary

Powerline RFI may be created on the line or picked up from a noisy source connected to the line. The RFI may reach the complainant via conduction along the line, by inductive coupling from the line or by direct radiation from the source (Figure 28). In some instances the RFI is a sum of all three modes of propagation. Inductive coupling may exist from one set of lines to another as well. RFI can travel many miles along power lines making noise source identification a time-consuming task.

Chapter 6

How the Power Company Locates Power Line RFI

While locating and correcting RFI emanating from the many devices in the neighborhood can be done by the amateur operator or the CBer, locating and pinpointing power line RFI is a different matter. The efficient antenna system made up of the overhead wires conducts and radiates a spot source of RFI very efficiently. Standing waves of RFI develop on a power line and produce spurious noise sources at intervals along the line that are exceedingly difficult to differentiate from the main noise source. (Figure 1).

In general, power companies do not recognize the efforts of non-qualified individuals in locating the source of power line noise. In my 16 years of experience in this matter, I have found only three radio amateurs out of thousands who had the expertise and general knowledge of power line characteristics to allow them to pinpoint a certain pole as a noise source.

Finally, the power poles are the property of the utility company and should not be climbed or pounded by an individual to determine a suspected source of RFI. In addition to being nonproductive, such practices are dangerous as a pole can pose a death trap to those persons not aware of the hazards involved.

This chapter, therefore, is intended only as an informational outline for the general public and is aimed at power company personnel interested in an overview of the subject.

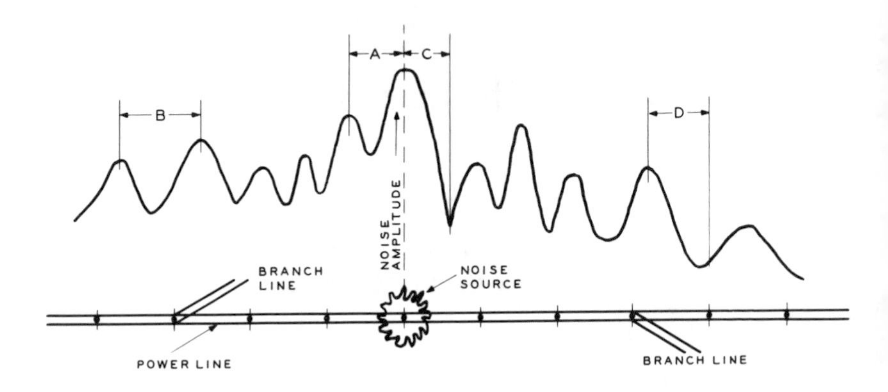

Fig. 1 "Standing waves" (a series of noise peaks and valleys) appear at intervals along the power line. As searcher moves along power line he will find out that the distance between noise peaks becomes less as he approaches the noise source (compare distance B with distance A). The distance between a peak and a valley also decreases as searcher nears the noise source (compare distance D with distance C). Variations in noise strength may be caused by radiation from branch lines, down leads, ground wires and nearby metallic objects. The distance between noise peaks and valleys usually decreases when the searcher makes his observations in the vhf region.

Locating Power Line RFI

The techniques of locating sources of power line RFI are similar to those used in locating devices creating RFI in a neighborhood. The interference must be traced to a particular pole (or poles) just as a noise-producing device in a residence or industrial area must be located before corrective measures can be taken.

If every source of RFI from overhead power lines had the same degree of intensity the sources would be less difficult to locate since it would then be possible to assume that the noise was a given distance from a particular complainant. Unfortunately, this is not the case. The intensity of power line RFI varies from day to day, even hour to hour, because of weather conditions, expansion and contraction of utility lines due to temperature changes, vibrations in conductors and the movement of power poles. The degree of difficulty in locating the source of noise on a particular pole, moreover, depends not only on the noise intensity but the frequency band at which the interference peaks. It takes a well-trained

RFI investigator with plenty of experience to quickly and accurately locate a source of power line RFI.

What You Can Do to Help Locate Power Line RFI

This chapter is divided into two parts. The first part outlines the general principles of tracking down power line RFI. The interested radio amateur, CB operator or layman can easily perform this task. The second part of the chapter concerns the pinpointing of the RFI by an RFI investigator--a skilled technician employed by the power utility.

When an untrained individual suspects the source of RFI, he should then give the location and the number of the pole to the local power company, along with a description of the noise and the time of day it was heard. (The pole number is commonly stamped on a metal plate attached to the pole).

Under no circumstances should the individual climb or strike the pole, shake the guy wires or otherwise disturb the installation. Power poles are not public property and indiscriminate tampering with high voltage circuits is dangerous, not only because of the chance of electric shock but also because the untrained individual can cause outage or damage for which he could be held responsible.

Most power companies have an interference locating program for their customers and advantage should be taken of this service. However, a little detective work by the individual to spot the general area of interference will help the trained investigator pinpoint the interfering pole or equipment with a minimum of effort. But do not assume the duties of the investigator!

The Initial Search

As discussed in the previous chapter, power line noise is commonly conducted down the lines and radiated by them as shown in Figure 2. The RFI radiated from a power line is confined to the general vicinity of the source, whereas RFI induced into the line (while weaker in intensity than the radiated field) can be carried along the power line for miles. This can exhibit confusing indications to an investigator unless he is aware of the problem. A radio amateur or CBer can often get an approximate "fix" on the RFI source with his beam antenna which will point in the general

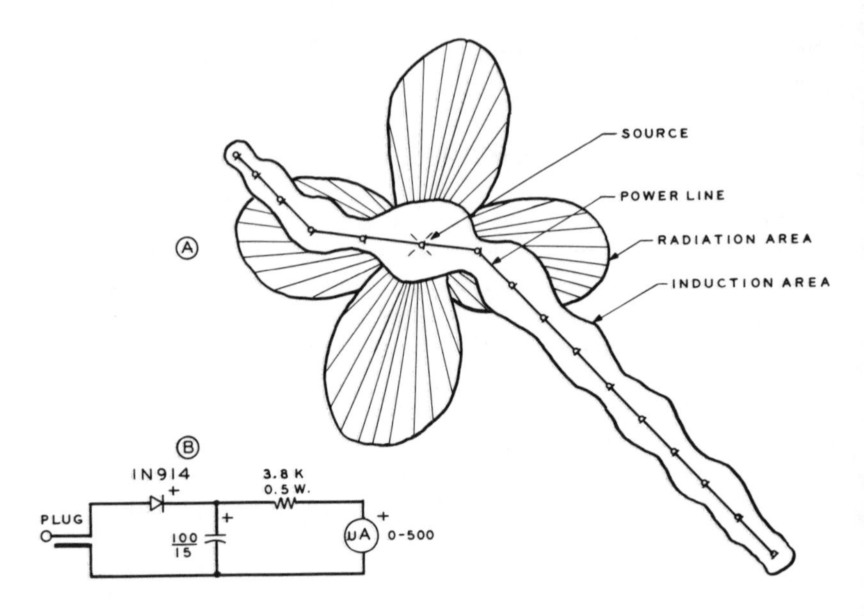

Fig. 2 Induction and radiation fields of power line RFI are shown in A. Radiation field is confined to the general vicinity of the noise source whereas the induced noise field can be carried along the line for miles. B-Simple audio level meter can be plugged into earphone jack of portable radio to provide visual indication of noise level.

direction of the radiation if it is turned to the loudest indication of noise in the receiver.

The noise searcher can be fooled by a second nearby source of noise radiation, or the initial search can consume a lot of time if there are two noise sources on the same power line. In any event, the searcher starts his noise hunt with the portable receiver in his auto, driving around in the general area where the RFI is loudest when his receiver is tuned to a clear spot at the highest frequency the noise is noted. Once the offending power line is located, the searcher drives along the line, taking general note of the noise situation. The RFI noise level will vary drastically as noise radiation occurs from down leads, ground wires and other pole hardware. The noise generally has peaks and minimums as the searcher proceeds along the power line; as he moves away from the RFI source the distance between adjacent noise peaks (or valleys) becomes greater. Conversely, as he moves closer to the noise source, the distance travelled between noise peaks or valleys becomes shorter. The intensity of the noise has no meaning at this time.

Zeroing-In on Power Line RFI

The final detective work is done in the vhf spectrum. When a vhf noise peak is reached, the searcher should continue moving along until the RFI is very weak. Then he should retrace his path back to where the interference was the loudest and continue on past it in the opposite direction until the noise is once again at a minimum. The searcher should drive along both sides of the power line, if possible, or around the block as more than one pole can be causing the problem.

Some amateurs and CBers believe they can stop RFI by hitting a power pole hard with a sledge-hammer. This is not true and such an attack can cause serious damage to the pole, the protective coating on the pole and the devices mounted on it.

The final thing to do is to note the pole number and location and inform the local utility company or agency. They will issue an order for corrective maintenance.

Training the RFI Investigator

The business of locating and eliminating RFI is a growing one and more and more utilities and communities are taking a serious interest in the problem. The RFI investigator fills an important need, and this service is a unique and important one, calling for an individual having the basic knowledge of power line construction (Figure 3) and radio communication--a rare combination. Individuals with a basic education in electronics can be trained in this new and interesting field.

The individual desirous of becoming a professional RFI investigator should have a basic knowledge of electricity equivalent to that provided by high school physics courses. As he will probably be a company representative who continually meets the public, he needs an aptitude for public relations; this aptitude is generally determined from pre-employment tests and personal evaluation. Above all, he must possess the ability to be tactful and get along with people.

In addition, the investigator must have initiative as he will be working by himself most of the time. He must be loyal to his employer and follow policy and procedures, yet show sympathy for the customer's problem. And he must not show anger when he encounters verbal abuse from a complainant, realizing that the abuse is only directed at him

Fig. 3 The source of RFI has been spotted by the investigator and he has summoned a lineman to place a RIV clip in between a string of bell insulators. Basic knowledge of power line construction and radio communication is an important combination required by the RFI investigator. Individuals with a basic education in electronics can be trained in this new and expanding field.

because he is the representative of the utility. When such a situation is encountered, the investigator must say to himself, "There is no use for both of us to be angry, one of us has to stay calm to solve the problem."

Above all, a very important trait of an investigator is his ability to *listen.* When he does talk, it should be across to the complainant as an equal, never down in tone.

The prospective RFI investigator is instructed that he will not be the first contact with the complainant. Due to the large area that an electric utility or agency serves, it is common for telephone calls to be routed to central areas for customer convenience. Thus, the complainant's first contact will be with an individual who knows absolutely nothing about RFI, but *does know* the information required to aid the investigator, such as time of interference and frequencies or channels affected; many utilities use a form which is filled out on initial complaint and passed to the investigator (Figure 4).

NOTICE OF INTERFERENCE ☐ Radio ☐ Television ☐ Amateur Radio ______ CALL LETTERS 72445

MAIL TO: *(See ESM 33.20.5)*

Street Address ______ Date ______ MONTH DAY YEAR

Name ______ Telephone No. ______

Nearest Cross Street ______ Someone is usually home at ______ AM ______ PM

City ______ District ______

How to locate if rural ______

NATURE OF INTERFERENCE

1. Customer has been troubled for ______ ☐ Hours ☐ Days, or ☐ Weeks
2. Trouble is experienced at ☐ all hours, ☐ occasionally, or state hour ______ AM ______ PM
3. Does it appear only when using electrical appliances? ☐ Yes ☐ No ☐ Do Not Know
4. Has equipment (radio/TV set) been checked by a serviceman? ☐ Yes ☐ No ☐ Do Not Know
5. Do neighbors have same trouble? ☐* Yes ☐ No ☐ Do Not Know

* If answer is yes, give name and address of neighbor ______

DESCRIBE FULLY NATURE OF TROUBLE ______

______ ISSUED BY ______ ISSUING OFFICE

Fig. 4 Representative "Notice of Interference" form. The form has three copies and is filled out on initial complaint to the utility company. Two copies of the form are given to the RFI investigator and one is held for reference.

The next contact with the complainant is a phone call from the investigator. At last the customer feels he is talking with an expert who will solve the problem!

What the Investigator Does

It is not a good policy for the investigator to make a specific time appointment with the complainant because the investigator, in planning his day's work, cannot foresee what he will encounter on his calls. There are times when an investigator is confronted with an RFI problem which cannot be solved immediately. In such a case, the best thing to do is to admit it to the complainant, tell him/her it is a unique problem and that additional help and time will be required. The investigator should ask for cooperation and patience and should stay in touch with the complainant.

Experience has shown the best trainee for this unique position, assuming he possesses the aforementioned qualities, is an ex-lineman. (It would also help if he were a radio amateur).

The reason the lineman is the first choice for an investigator is that he

Fig. 5 Banks of capacitors on power poles are rapidly being replaced because of environmental contamination caused by leaks. Poles are being marked that have contaminating capacitors mounted on them. Linemen should be instructed in the problem of handling capacitor leaks.

knows the construction details of overhead power lines. On the other hand, if a potential investigator is at hand, the utility or agency must train him in the various aspects of overhead power lines.

First, the trainee should be taught, by sound recordings, the difference between the various sounds of RFI as heard on a receiver. This is important as he will be tracing the sources of RFI largely by ear. His training should also include the visual characteristics of interference as seen on television screen. In addition, he should have a good working knowledge of a tv receiver as he may have to discuss receiver filters with the complainant or serviceman, or determine if the interference is originating within the tv set itself. The trainee should be given a brochure illustrating the different patterns of TVI and he should familiarize himself with these patterns. The brochure will be of great help in working with complainants, as pictures of various forms of interference are useful in explaining the problem to nontechnical individuals.

More Training for the RFI Investigator

The next phase of his training is to assign the would-be investigator to an overhead line construction crew. This is important as he will learn the construction techniques and the jargon used on the job. Incorrect terminology which shows a lack of basic knowledge would severely reduce his influence with the very workers who will eventually respond to his written request for work on an affected line.

In the time the trainee is assigned to a construction crew he is expected to learn the various items installed on a pole. He learns the difference between an eyebolt and a hog eye, and understands what the hardware is and what it does: the fuseholder, the cutout, the lightning arresters, the pothead, the capacitor banks and so on (Figure 5).

The trainee also learns the difference between open-wire telephone line and power circuits, otherwise he'll find it embarrassing when a pole carrying a telephone line is turned in as a pole carrying electric lines!

The trainee must be instructed in the various causes of spark discharge from power lines and the methods used to correct this nuisance. When it is felt the trainee has absorbed most of the instruction given so far, he should be assigned to a senior RFI investigator, working in a prime television area. The sources of RFI from overhead power lines and from other sources are a complex mix and provide excellent on-the-job training for the beginning investigator.

The Investigator "Graduates"

When the trainee has become thoroughly familiar with the equipment and methods used by the RFI investigator, he is ready to trace sources of RFI on his own, under supervision. He makes contact with the complainant by telephone and in person. This is the point at which the trainee must develop his "cool" with regard to the customer and yet be able to admit to his instructor that he might not know the solution to a knotty problem. Mistakes made by the trainee are corrected by constructive criticism from the experienced investigator and made when they are out of hearing of the complainant.

Finally, when the trainee appears to "know his onions", he is assigned to an investigator working in a fringe television area where TV signals are weaker and more affected by RFI, thus making cures more difficult.

On-the-job Training

A vital phase of training (carried out during all the previously mentioned instruction) is tracing RFI for radio amateurs and CB operators. In addition to learning the terms and words used by the power line construction forces, the trainee needs to know the unusual jargon used by these groups of individuals. He must learn the frequencies assigned to amateur and CB operators and the relationship between frequency and wavelength. This knowledge can be partially picked up on the job and can also be learned from various handbooks published exclusively for hams and CBers.

The trainee should also know that the beam antennas used by many hams and CBers possess signal gain that the investigative instruments he uses do not. Thus these big antennas have the ability to pull in low-level radio noise from the power lines over considerable distances and noise objectionable to a ham or CBer may not be objectionable to a casual tv viewer next door. Tracing sources of power line noise for this group of customers is often difficult, especially when the trainee realizes that he must eventually pinpoint the exact location of the RFI source.

RFI Search Equipment

If a company spends the time and effort to train an RFI investigator then it also must supply proper receiving equipment for the investigator's vehicle. The equipment should include a shortwave receiver with a vhf converter, plus a portable am-fm radio and portable tv receiver.

In addition, the vehicle must be equipped with a shielded ignition system. An extra item of value is a small hand-held Yagi beam antenna cut to tv channel 6 or 7 which can often indicate the direction of television interference. It is a matter of judgement and economics as to the equipment selected. The better the investigator's training and equipment, the lower the overall cost of tracing RFI.

The RFI Assignment

The time comes when the supervisor and instructor are convinced that the newly-trained RFI investigator is capable of being assigned to an area.

The investigator's first duty, when assigned to an area, is to introduce himself to the supervisors of overhead construction and the linemen who will correct power line problems that he locates. At this time, he should request that a crew be made available to work with him when he locates an RFI source on a pole. By doing this, the crew will hear the sound of RFI and know what they did to correct the fault.

Once the investigator has gained the confidence of the crew and shown his ability to locate the precise pole causing the difficulty, he can go on his way, knowing that the crews will do their utmost to perform the corrective maintenance he recommends.

If this procedure is not followed and the investigator is unlucky enough to pick out a pole that does not correct the RFI after it is worked on, and then turns in another pole on a subsequent job order, he can be sure his competence will be discussed at length by the construction supervisors and his superior. If too many mistakes are made, the supervisors will demand the investigator's presence every time a crew goes out to clear an RFI problem.

One important fact in RFI-tracking must be explained to the construction forces: it is possible for one source of RFI to be masking another source in the same vicinity, and that a return to the same area may be necessary to further reduce the RFI. Of course, if time permits, it is a good idea to check all around the area after a cure is achieved as this may make a second trip unnecessary. In all cases, the investigator owes it to himself to become thoroughly familiar with the power lines in his assigned area. This saves time when problems arise.

Cable TV

Television signals cannot reach some areas either because the signals are blocked by high surrounding terrain or because the distance from the transmitter is too great and signals are extremely weak. If there is a sufficient number of people in such a fringe area, a *community antenna* consisting of a high tower, sensitive receiver and amplifiers can be utilized, the resulting signal "piped" to viewers over a coaxial line.

Because the cable tv system is dealing with a weak incomingsignal it is especially susceptible to RFI problems. The RFI investigator is in a delicate situation because the combination of weak signal and high gain antennas allow interfering noise as far removed as 15 miles to be picked up. And within the 15 mile distance, multiple noise sources may be located. The economic cost of finding and curing all the sources can easily be beyond the resources of the utility. This can create animosity between the cable tv organization, the tv viewers and the utility.

On the Trail of Power Line RFI

A complaint is received by the power company about RFI. The caller is questioned about the characteristics of the interference, and the frequencies or channels involved. The investigator reads the report and calls back with more specific questions. If, in answer to a question, the customer states there is a burst of dot-dash lines, or shot lines, on the tv screen which appear for a particular time interval, the smart investigator asks the complainant to disconnect all household appliances by pulling the power cord, unscrewing the fuse or tripping the circuit breaker. If this simple step eliminates the RFI, the cause can be quickly determined and the problem solved by telephone rather than by a more costly trip to the site of interference. An examination of past records of a local utility shows that this preliminary step cures the RFI problem better than 20 percent of the time.

If the cure cannot be effected by phone, the investigator must visit the site. He turns on his receiving equipment about a mile from the residence as there is a good chance the source of interference may be picked up enroute, thus saving later driving about to pinpoint the source.

The investigator arrives at the residence and after an appropriate introduction requests to see or hear the interference. He checks to see if the RFI sounds the same as that picked up when he entered the area. If so, the location of the source is half-completed.

If the investigator suspects the RFI source of being out of the residence, he must begin to trace it with his receiver tuned to the highest frequency on which the noise can be heard. As he walks or drives away from the residence, he notes if the interference becomes weaker or stronger and observes if it has "standing waves" of maximum noise. (See Figure 1).

This term refers to the fact that RFI usually shows regular peaks and valleys of intensity as the investigator moves along the power line. If the noise peaks tend to get weaker as the search progresses, it is good indication that the wrong direction has been chosen. The investigator turns around and proceeds in another direction--possibly in several directions--before the next noise peak is observed.

The investigator now drives or walks along the suspected power line, which in many cases is located along rear property lines. He listens to the noise and watches the signal strength meter for an increase in intensity. When an increase is noted, he changes frequency to a higher range, or band. Familiarity with his equipment and with the "signature" (unique identifying sound) of power line noise combine to tell the investigator he is heading in the right direction.

When the standing waves of noise become closer together insofar as distance is concerned, and the noise peaks grow in intensity, the investigator knows he is approaching the source.

When a noise peak is reached in the vhf spectrum the investigator continues ahead slowly until he notices the RFI starts to drop in intensity. He then reverses course and retraces his path to the noise peak. This could be between two poles and if a directional antenna is part of his equipment, a definite pinpoint of the source can be quickly made. If the source remains indeterminate, the exact location may have to be made by *gently* tapping the suspected poles with a sledge hammer.

It takes only a slight rap to disturb the RFI creating components on the pole. As already stated, some amateurs and CBers believe they can stop power line RFI by hitting the pole hard with a sledge hammer. This is *not* true and can cause damage to the wires and devices on the pole.

Once the investigator has found the source of the interference, he notes the pole number and location and issues a Job Order for corrective maintenance which should include tightening the hardware, bonding, staples, tie wires, clamp tops and--if dead ends are present--the installation of the RIV clip previously mentioned.

When Tapping the Power Pole Fails

All of this sounds simple--but often it is not. When there is a loose metal-to-metal contact on a pole, a light tap on the pole changes the sound of the RFI or may shut it off for a short time. But suppose tapping the suspected pole shows no change in the RFI? Is this the right pole? Unfortunately, there are devices on a pole which might cause interference where the investigator cannot alter the characteristic noise of the RFI. Such devices include lightning arresters, transformers, cracked insulators and a broken bond between two washers. A single clue may point out these items if the pole-tapping technique fails: if the RFI persists *during and after rain* especially in areas where rainfall is limited to certain times of the year.

The lightning arrester. The lightning arrester is connected between a primary conductor and ground to provide a return path when lightning strikes the circuit, thus protecting the equipment (Figure 6). The arrester is a porcelain cylinder containing an element which limits the flow of current to ground under normal circumstances. The element is called a *grading resistor.* During a power surge caused by lightning the char-

Fig. 6 The lightning arrester is connected between conductor and ground to provide a return path when lightning strikes the circuit. Arrester contains special resistor that allows voltage surge to pass to ground.

acteristics of the resistor change and allow the surge to pass safely to ground. After a series of lightning strikes, the resistor may develop a continuous corona discharge which can cause severe RFI. Tapping the pole will not affect this discharge.

The audible buzzing noise created by the corona is very high and may be heard or noted on a sonic locator (Figure 7). The locator is placed parallel to the arrestor ground lead for best identification of noise.

When arresters are suspect a crew must be called to disconnect them from the primary conductors. The investigator listens to the noise on his receiver: the first arrester is disconnected--no change in the noise level. Is he a hero or a goat? The second arrester is disconnected--a slight drop in noise is noted. When the third arrester is disconnected the RFI disappears and the investigator is a hero at last. The defective units must now be replaced.

The transformer. The pole-mounted transformer (Figure 8) is a prime suspect if the RFI noise peak *drops off* rapidly above 13 MHz, or if the

Fig. 7 Corona discharge may be picked up by sonic locator which is sensitive to frequencies in the 30 to 50 kHz region.

Fig. 8 Pole-mounted transformer can cause RFI because of insulation leakage or intermittent winding. Transformer replacement is expensive and time consuming and circuits must be de-energized during work.

Fig. 9 Ground-mounted transformer can be a serious source of RFI.

noise starts at about 13 MHz and can be heard up into the vhf spectrum. When a transformer is banked with other transformers' secondary systems, the secondary taps must first be removed and the primary side de-energized for the RFI to stop. The RFI investigator is reluctant to disconnect a transformer until all other possible sources of RFI have been investigated. Replacing a transformer is an expensive and time-consuming task and the investigator never issues a job order unless all other possibilities have been exhausted.

In the case of a cracked insulator or broken bond wire between washers, the investigator must be present to note any change in RFI level as the conductor and bond wire are lifted from the pin and insulator. All other bonds have to be disconnected from the related hardware before the broken bond can be located. A nearby radio receiver will quickly pinpoint the problem when the defective device is located.

Underground and Pad-mounted Transformers

The most difficult noise source to locate is one coming from an underground transformer, or one that is ground-mounted on a concrete pad (Figure 9). RFI from these sources rarely affects a television receiver, the major noise frequencies being below approximately 9 MHz, a frequency range not conducive to pinpointing the source. Noise will be noted at underground riser poles which can lead the investigator to jump to the conclusion that the RFI is caused by potheads or lightning

Fig. 10 Model 700 portable interference locator covers broadcast band through tv channel 13. Receiver is battery equipped and has video output terminal so photograph of interference can be made with auxiliary oscilloscope.

arresters. Similar false noise peaks can be observed at corner poles or switches.

In an underground tract, the distribution transformers can be switched to be served from one of two directions by the same circuit. Each transformer has two primary switches which are closed so as to serve the next transformer. The last transformer will have one switch open to avoid a loop feed in the circuit. An elaborate switching procedure must be followed in order to locate the source of RFI and the resulting power outages inconvenience the customers on the circuit. To date, these transformers have not been a great source of RFI but when they are it can be a very expensive problem to correct, especially if the transformer is inconveniently located and a crane with a long boom must be used to lift the transformer to the street.

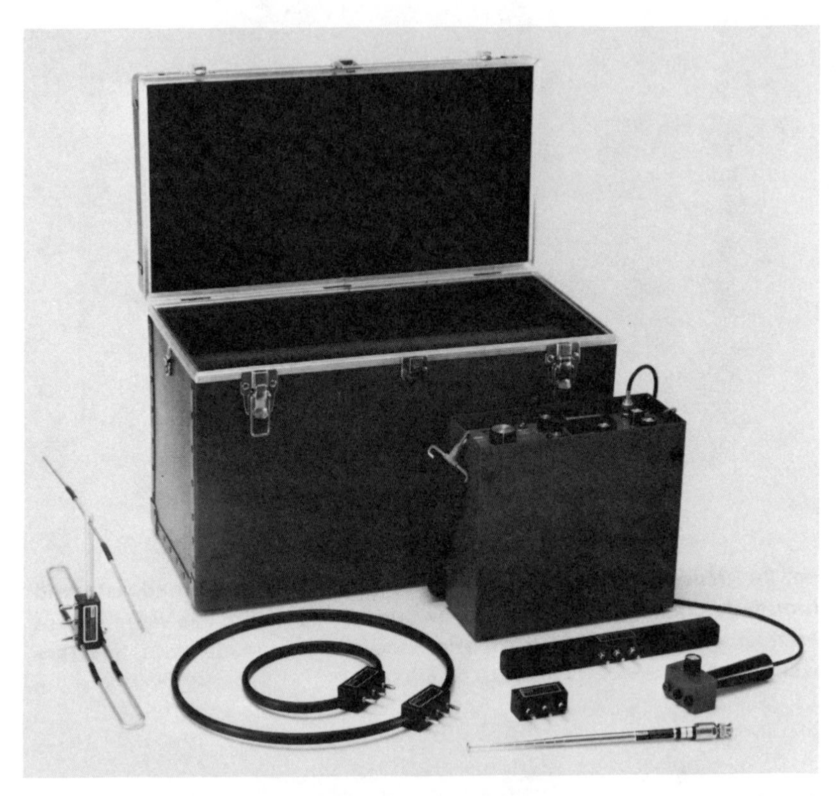

Fig. 11 Model 700 locator has assortment of plug-in antennas and can be purchased with carrying case and accessories.

RFI Location Equipment

Shown in Figures 10 and 11 is portable interference location equipment used by some utility companies. The Model 700 locator is a tunable receiver covering broadcast through tv channel 13 which has an assortment of plug-in antennas for direction-finding a noise source. The receiver is battery operated and has a video output terminal so that a photograph of the interference may be made with an auxiliary oscilloscope.

Fig. 12 RFI "gun" is useful in tracking vhf noise. Compact receiver and beam antenna can pinpoint noise source from a considerable distance.

The "RFI Gun" shown in Figure 12 is a wideband receiver covering channel 13 through the uhf region. It has a built-in broadband directive antenna and is battery operated. A simple, miniaturized version of this receiver is shown in Figure 13. The "Gun" is useful in locating the source of interference on a particular circuit on a pole. These items are available from Micro-Tec Co., 703 Plantation St., Worcester, MA 01605.

RFI Reporting Forms

It is helpful to have a standardized RFI reporting form and a RFI job order form. Sample forms used by a large utility company are shown in Figures 4 and 14. Required information for the form includes: Name, address, location and telephone number of complainant; nature and time of interference and a space for describing the problem. On the back of the form is a check list covering investigation and completion of the report. This check list includes:

Called customer
Visited customer
Interference found

Fig. 13 Hand held RFI "snooper" picks up RFI and aids in locating noise source. It is designed to be used in vicinity of high voltage lines.

Could not locate interference
Found possible source
Customer created problem
Issued work order
Received completed work order
Interference has cleared
Advised to re-call
Date and time

RADIO & TELEVISION INTERFERENCE JOB ORDER

TO: OPERATIONS SUPERVISOR OR SUPERINTENDENT

02226

OF ______ (DISTRICT OR DIVISION)

CIRCUIT NAME

NEAREST CROSS STREET

CITY

CUSTOMERS ADDRESS

CUSTOMERS NAME

OD 415 NO.

POLE NO.	KV	LOCATION	WORK CODE

DATE ISSUED

MANAGER OF TELECOMMUNICATIONS APPROVAL

RETURN ON COMPLETION TO:

☐ SAME ☐ OTHER ______ (PHONE)

Fig. 14 Job Order Form for RFI complaint. Four part form authorizes work to be done after complaint is filed. Front of form outlines work to be done and rear of form summarizes work completed by line crew.

RADIO & TELEVISION INTERFERENCE JOB ORDER RECOMMENDED CODES

Code A – Tighten all hardware, bonding, and connections
Code B – Install bonded type insulators
Code C – Install post type dead end insulators
Code D – Install bonding
Code E – Remove slack from span
Code F – Disconnect lightning arresters for interference check
Code G – Install R.I.V. clips in dead ends
Code H – Further Remarks ______

DETAILED DESCRIPTION OF WORK DONE

MANHOURS

DATE COMPLETED

SIGNED

TITLE

The "Radio and Television Interference Job Order" is a four-part form used within the utility. The face of the form includes the complainant's name and address and spaces to fill in the pole number, line voltage, location and work code (for the time charged to the job). The date issued is also included as well as a sign-off by the manager of the project.

On the reverse of the form is space for description of the work done with a check list for action which includes-

Tighten all hardware, bonding and connections
Install bonded type insulators
Install post type dead-end insulators
Install bonding
Remove slack from span
Disconnect lightning arresters for interference check
Install RIV clips in dead ends

RFI Follow-up

Visits by one of the authors of this handbook to various utilities have shown that some of them ignore the necessity to provide at least 1½ inches clearance between unrelated metal parts on poles and crossarms (see items 3 and 4, Fig. 22, page 91). Much pole-generated RFI can be traced to lack of sufficient clearance in this, and similar, areas. It is important that hardware associated with conductors be clear of crossarm braces, guy wires, ground wires, etc. Attention should be given to clearance between crossarm braces and washers backing up conductor hardware bolts. In some instances it may be necessary to provide longer crossarms to achieve the desired clearance.

Chapter 7

Noise Reducing Bridges for Your HF Receiver

What can be done to reduce RFI at the receiver? If the noise source cannot be located, or the cure too expensive or time consuming, what can the listener do?

Some communication receivers have built-in noise clipping circuits or other devices that "punch a hole" in the noise. Good noise reduction circuits are sophisticated and expensive and most simple circuits are ineffective. In addition, such circuits must be tailored to match the equipment in such a way that general information on adapting noise suppression circuitry to a particular receiver is of questionable help.

Experiments have shown, however, that a large amount of random noise can be phased out before the antenna terminals of the receiver. The noise is picked up on a "noise antenna" situated so that it receives a maximum amount of noise and a minimum amount of wanted signal. Position of the noise pickup antenna is determined by experiment. In many cases, twenty or thirty feet of wire running parallel to a nearby power line and a safe distance away from it will suffice. The noise from this antenna is blended in a bridge circuit with the noise from the regular station antenna in such a way that the noise signals are in opposition, permitting the desired signal to be heard relatively noise-free.

A Good Ground Connection is Required

Whether or not a noise bridge is used, a good ground connection to the receiver will reduce the noise pickup that feeds from the line directly into the input circuits of the receiver (Figure 1). You will have to experiment

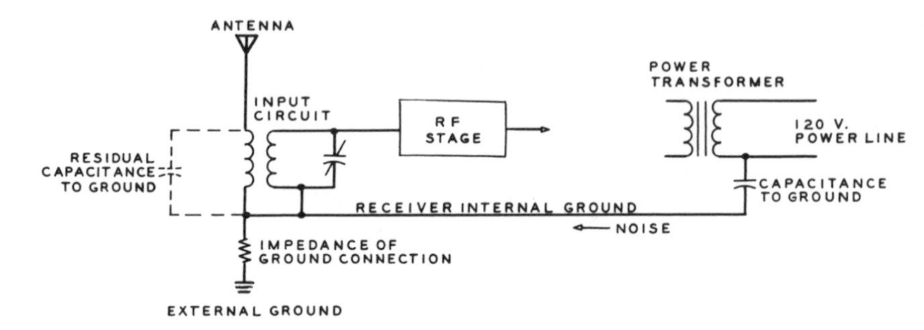

Fig. 1 Noise may be coupled from the power line into ground circuit of your receiver. A good external ground connection bypasses the noise back to ground. Ground wire should be short as possible.

with ground connections to find one that provides the least noise pickup. The ground wire should be as short as possible as a long connection may act as an antenna and actually increase the noise level in the receiver.

The electrical conductivity of the earth varies greatly from one location to another. Moist, rich soil makes an excellent ground while dry, sandy soil with low mineral content makes a relatively poor ground. In any case, several ground rods driven into the soil and connected in parallel will do a better job than a single rod.

The Cold Water Pipe

A connection to a cold water pipe in the home may be the most effective ground that many radio amateurs and CBers can easily obtain. Such a ground connection can be suprisingly effective if the plumbing system is made of soldered copper pipes. Some newer homes, however, have a plastic section of pipe that connects the home water system to the city water main. This effectively isolates the water pipes from ground.

Older homes having galvanized iron water pipes often provide a poor ground because the pipe sections are electrically insulated from each other by the joint sealant used to prevent leakage. In both of these examples, the best place to make a ground connection is to the metal water pipe on the street side of the water meter. The ground connection can be improved by connecting the water pipe to one or more ground rods driven in the earth outside the residence.

A good ground connection in many homes can be made to the air conditioning and heating ducts running beneath the floor. Sometimes a

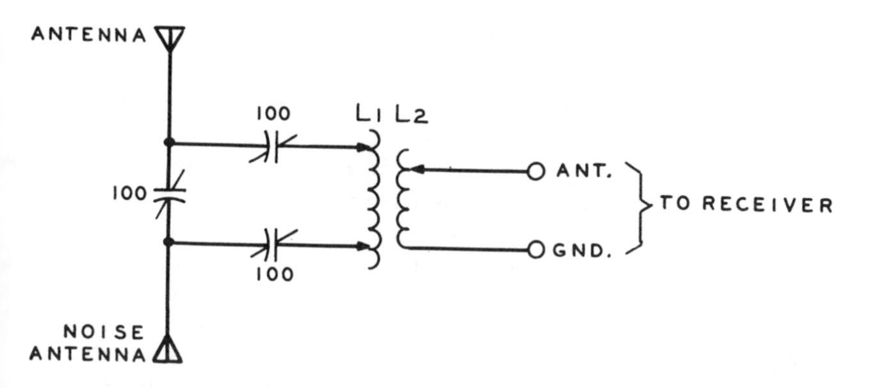

Fig. 2 Simple noise balancing bridge. Coil L1 is ten turns, 1-inch in diameter and 2 inches long. Coil L2 is five turns, closewound around center of L1. Taps and capacitors are adjusted for minimum noise.

jumper between a heating duct and a nearby water pipe will provide an effective ground.

Outside the home, the lawn sprinkling system made of copper pipe or galvanized iron sections can provide a good ground system. The station ground lead can be attached to the copper pipe with a clamp. Soldering is ineffective if there is water in the system as it will be impossible to heat the pipe sufficiently to make a good joint. As a last resort, if the sprinkling system is ineffective, a random length of insulated copper or aluminum wire may be buried and used as a ground connection. The wire should be buried about a foot below the surface of the earth.

The Ground Rod

A good ground rod may be purchased at any large electrical supply house. The rod is made of steel with a copper outer plating. Rods come in various lengths and a four to eight foot long rod will do the job. Several short rods driven in the ground about a foot apart are easier to install than one long rod, particularly in hard, rocky soil. The rods can be connected together by means of copper plated pipe strap. A good connection from the rods to the receiver can be made of the outer braid removed from a length of old RG-8/U coaxial cable.

The Noise-Balancing Antenna Bridge

A simple version of the noise balancing bridge is shown in Figure 2. It may be used with any high frequency communication receiver and when properly adjusted will provide over 15 dB rejection of power line noise while dropping the desired signal less than 3 dB. Because of variations between receivers, setting of the coil taps will have to be done by experiment. The taps and capacitor settings are adjusted to balance out the power line noise while retaining maximum signal strength.

A more effective noise balancing antenna bridge is shown in Figure 3. A ferrite core wideband transformer is used with the noise antenna connected to one winding and the station antenna connected to a second winding. The windings are connected out of phase. The resulting net signal is taken from a third winding which is connected to a balun (balance-to-unbalance) transformer. Ideally, if the signal in the noise winding is equal in amplitude and phase to the noise component of the signal in the antenna winding, only the signal component will be found in the third winding which is coupled to the station receiver through the balun.

This exact balance of signal and noise is unlikely to occur so additional controls are added to adjust the amplitude of noise components from each antenna to equalize each other in the output winding of the transformer.

Constructing the Noise-Balancing Bridge

The components of the bridge are mounted in a small aluminum box approximately 6" x 3" x 2" in size. Capacitor C1 must be insulated from the box with a shaft extension made of insulated rod to reduce hand capacity during adjustment. The capacitor is supported by an insulating bracket bolted to the box. The coil L1 can be supported between a capacitor terminal and a phenolic tie-point strip. A small clip is used to tap the coil. It will be easier to attach the clip to the coil if every other turn is indented (pushed in 1/8-inch) to provide more space for the clip.

Transformers T1 and T2 are wound on small ferrite toroid forms (Q2 material). Transformer T1 is trifilar wound, that is, three windings are wound on the core in parallel as one winding. Before the wires are wound on the core, they are twisted together at one end and held in a vise. The other ends are temporarily twisted together and anchored in the chuck of a hand drill. Give the drill a tug to take out the kinks in the wire and then wind the wires with the drill, twisting them to about four twists per inch.

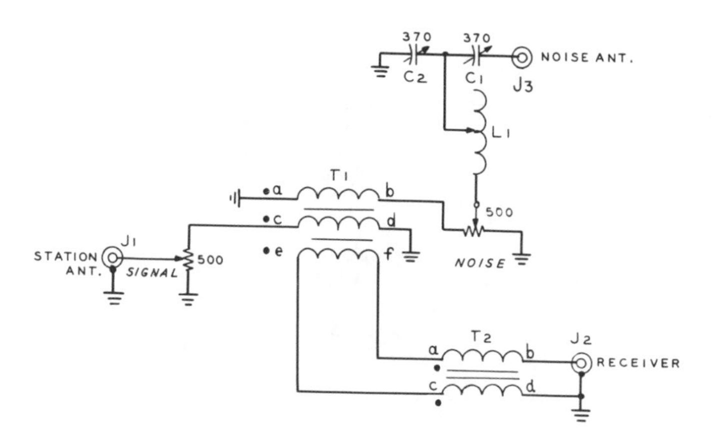

Fig. 3 Noise balancing bridge. C1, C2: 370 pF compression-type trimmer with shaft (air capacitors may be used). L1: 45 turns No. 22 enamel wire, 1-inch in diameter, turns spaced wire diameter. T1: 16 turns No. 18, trifilar wound on 3/4-inch diameter ferrite core, Q2 material (permeability of 125). J.W. Miller F-125-2 or Amidon FT-82-61. T2: Same core material, 18 turns No. 18 bifilar wound. Note: dots on drawing indicate adjacent ends of windings.

The two wires of balun T2 are twisted in the same manner. When completed, the wire skeins are wound on the cores, spacing the turns neatly so as to fill out the core area.

The last step is to strip the insulation from the ends of the windings and, with an ohmmeter, locate the respective wires *ab, cd,* and *ef.* Join *a* to *d,* pair off *e* and *f,* which leaves *b* and *c.* The bridge is then wired as in the illustration.

Using the Noise-Balancing Bridge

A good level of noise should be established to aid in the initial balance of the bridge. The two antennas are connected and the potentiometers set for maximum noise signal (control R1) and a low level of desired signal (control R2). Next, adjust capacitors C1 and C2 and the tap on coil L1 for maximum noise. The bridge is now ready to balance.

First, advance control R2 (signal) to maximum and control R1 (noise) to minimum. Now, slowly rotate R1 from the minimum setting to a point where the noise drops in strength. Alternately adjust R1 and R2 for best noise null, trying to achieve this with R2 near its maximum position for

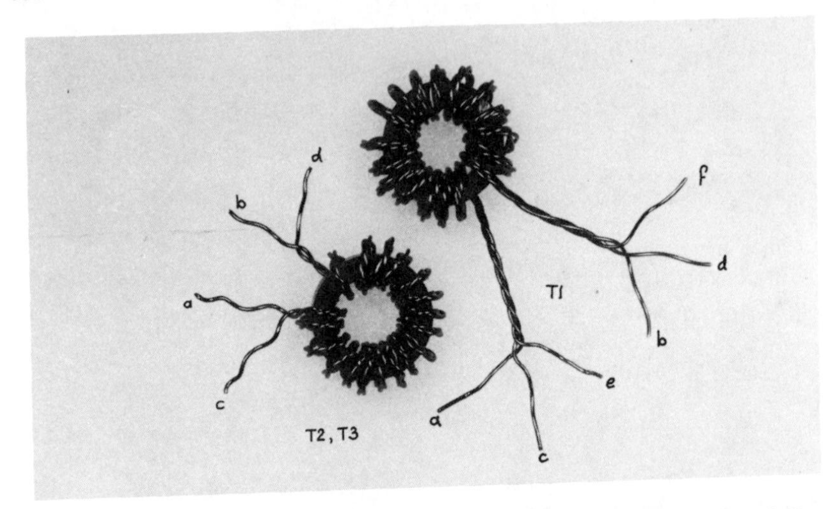

Fig. 4 The transformers used in the noise bridge. If the ends of the windings are identified with a drop of paint before the skeins are wound on the core it will be easy to make the proper connections to the circuit.

best signal strength. Some final adjustment of C1 and C2 may be helpful.

This circuit has been used at vhf by eliminating the network composed of C1, C2 and L1 and connecting the arm of the noise potentiometer directly to a small three element beam aimed at the noise source. During periods when the power line noise reached S9 it was possible to reduce it sufficiently to restore an otherwise unusable band.

(This Antenna Noise-Balancing Bridge was designed by VK3XU and described in "Amateur Radio", the publication of the *Wireless Institute of Australia*).

Chapter 8

RFI From Nonlinear Devices, Snivets, Barkhausens And Other Mysterious Sources

Nature can be an annoying source of RFI when aided by strong rf signals from nearby broadcast, shortwave and vhf fm and tv transmitters located near the larger cities. And home entertainment equipment can mysteriously generate RFI that is very difficult to locate. One particularly annoying type of interference is that caused by harmonic radiation from a nonlinear source in the presence of a strong rf field. A rectifier is a common nonlinear source as it passes current readily in one direction and opposes the flow of current in the opposite direction (Figure 1). This action produces a signal wave that is rich in harmonics of the original signal.

Many rectifiers exist in nature, the most common of which are oxides (rust) and other corrosion products of metals. The efficiency of these natural rectifiers is low but some products such as lead sulphide (galena), silicon and germanium can form excellent rectifiers. Copper oxide is a good rectifier and at one time was used extensively for power rectification.

The rectifying ability of corrosion byproducts is well known by engineers and if such objects are in a strong rf field, harmonics of a signal (or signals) will be created and radiated. The strength of the harmonics is a function of the rectifier efficiency, the strength of the signal and the "antenna effect" of the conductor system coupled to the rectifier.

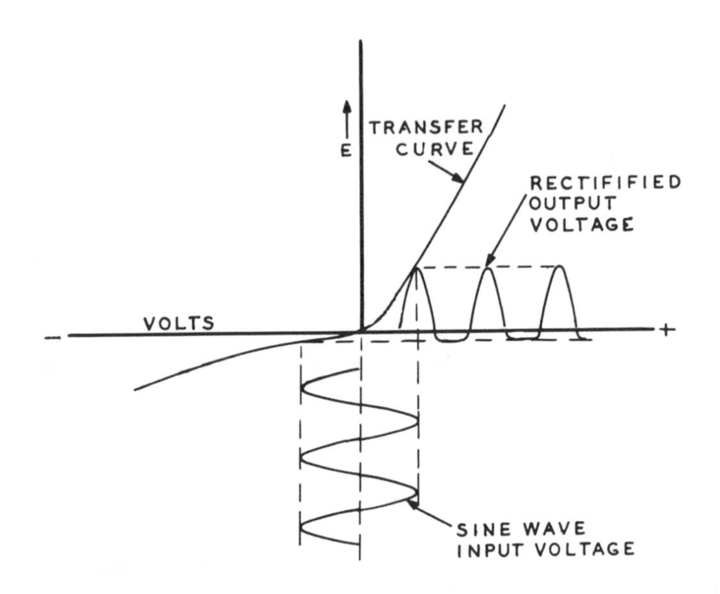

Fig. 1 Diode rectifier is a nonlinear source that provides many harmonics of a single signal. Output waveform is rich in harmonics. Many diodes are found in Nature that contribute to RFI.

Nonlinear Systems

Nonlinear devices which produce rf harmonics are present in most neighborhoods. Rusty or corroded joints in pipes, poor electrical connections, bad antenna joints and the usual maze of ducts, pipes, fixtures and wires found in homes, offices, stores and apartment buildings are the worst offenders. Corroded joints that can act as rectifiers may be found in the following:

Air ducts
Electrical conduit and BX cable
Electrical wiring (especially aluminum wiring)
Fences
Furnaces
Gutters and drain pipes
Guy Wires
Hot water pipes and heaters
Lightning arresters
Radiators and hot air registers
Radio and tv antennas and receivers
Sheet metal roof and walls
Stove pipes
Switch boxes, fuses and outlets
Telephone lines and equipment
Utility lines
Water and sewage pipes

All of these objects, and more, can produce signal rectification which shows up as harmonics of the original signal, or as cross modulation and mixing products of two signals. All of these actions can produce TVI and RFI.

Rectification, Mixing and Cross Modulation

Rectification creates spurious signals in a nonlinear device. *Mixing* combines two signals to produce sum and difference frequencies. For example, if a 900 kHz signal is mixed with a 2300 kHz signal, mixing products occur at 3200 kHz (the sum frequency) and 1400 kHz (the difference frequency) as well as the two original frequencies. Thus, new signals are generated that are seemingly unrelated to the frequency of the mixed signals.

Cross modulation is the transfer of intelligence from an unwanted signal to a wanted, weaker one. If a receiver, for example, is tuned to a wanted signal at 3.9 MHz and a very strong broadcast signal is at 1.4 MHz, the modulation of the strong signal is imposed on the wanted signal even though the broadcast signal is well outside the passband of the receiver. Cross modulation can be produced in any nonlinear device external to, but near, a receiver. Multiple signals can produce multiple cross modulation.

External cross modulation commonly occurs in the vicinity of large areas of rusty metals in intimate contact that are not bonded together. To cure this problem, it is necessary to locate the objects and electrically separate them or bond them together. This is often easier said than done. A good example of signal rectification and reradiation is the story of "Radio 98".

The Case of Radio 98

It has been said that if something can go wrong, it will. And this happened at Radio 98, a broadcaster operating on 980 kHz. It was not necessary to trace this case of RFI because amateur radio operators living as far as 15 miles from Radio 98's transmitter could identify the loud signal in the middle of the 80 meter band at 3920 kHz. It was radio 98 for sure, and complaints poured in to the station engineer. It was the fourth harmonic of the station that was causing the problem, the amateurs said, because 980 kHz x 4 = 3920 kHz. The worried engineer replied that the station had been carefully checked and there was no evidence of excessive harmonic radiation which, in any case, could not travel 15 miles and cause severe interference at that distance. Finally he suggested that the hams check their receivers.

Given this polite brush-off, the amateurs contacted the local power company for assistance. The RFI investigator was alerted to the problem and, even though it did not seem to involve the power company, a search for the interference was instigated as a measure of good will.

The investigator tuned his mobile receiver to 3920 kHz and drove south along a freeway, listening to the spurious broadcast signal. As he approached the transmitter of Radio 98 the signal became stronger. He lost it about five miles from the station but picked it up after travelling another mile and then heard the signal for about six more miles along the freeway.

Next, he entered another freeway heading west and drove another 10 miles, hearing the station intermittently on 3920 kHz. Finally, the investigator drove back to Radio 98 and invited the station engineer to take a short ride with him. The engineer accepted and after a few miles of driving and listening was convinced that something was wrong. The problem was: what was causing the interference and how to find it?

The engineer began a close examination of the transmitter and antenna system. The next day he phoned the investigator and told him he had found the trouble. One of the guy wires on the station tower had a leaky insulator plus a splice in the middle of the wire. By chance, the wire was self-resonant at about 3920 kHz. In the strong rf field of the antenna the insulator had arced at some time and a charred area was created across it. In fact, the music of Radio 98 could be heard in the tiny spark. To make matters worse, the splice in the guy wire had acted as a rectifier for the signal of Radio 98 and enhanced the fourth harmonic energy of the spark in a subtle manner. Replacement of the insulator and the guy caused the signal of Radio 98 to disappear from the 80 meter band, much to the relief of all of the local radio amateurs.

The Case of the Mysterious 160 Meter Signal

The RFI investigator received a frantic telephone call from one of the local amateurs. "The radio contest is coming this weekend! And there's a jumble of broadcast signals in the 160 meter band. What can I do?"

That evening the investigator drove over to his friend's house. Sure enough, as he pulled up in front of the house he heard a mixture of voice and music on his portable battery radio tuned to the 160 meter band. The worried amateur pleaded with the investigator to locate the problem as the operator's contest started in a few hours.

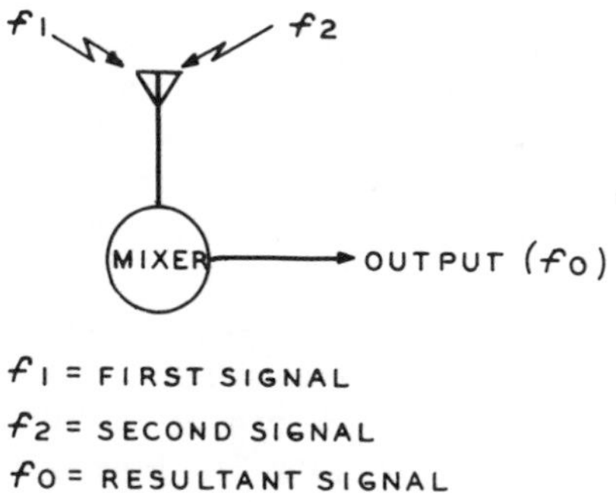

f_1 f_2

MIXER → OUTPUT (f_0)

f_1 = FIRST SIGNAL
f_2 = SECOND SIGNAL
f_0 = RESULTANT SIGNAL

Output signals are fo, f1, f2, fo-f1, fo + f1, fo + 2f1, fo-2f1, fo + 3f1, fo-3f1, fo-f2, fo + f2, fo + 2f2, fo-2f2, fo + 3f2, fo-3f2 and so on

Fig. 2 Sum and difference products of signals and harmonics produce a baffling mixture of spurious signals that cause severe interference.

The investigator knew that overhead power facilities could create cross modulation RFI. Under normal circumstances he would assign two junior investigators and a two-man maintenance crew to locate the source. They would check for guy wires touching each other, check pole and crossarm hardware, and check the neighborhood in the complainant's area for loose rain gutters, loose roof flashing, rusty tv antenna guy wires, scraping service entrance conduit and the ham operator's tower and beam. But there was no time for such an extensive survey. The investigator was on his own.

He entered the ham's house with his portable radio tuned to 1820 kHz and went over the entire house, holding the radio high in the air and then close to the floor. All the electrical appliances were disconnected, one by one. The burners of the gas stove were turned on and off. The door to the refrigerator was opened and closed. Nothing could be found to stop the RFI.

The attic area was checked and finally it was decided that the underneath area of the house would have to be inspected. The investigator crawled beneath the house on his hands and knees, bringing the radio with him. He rolled over on his back, keeping an eye out for black widow spiders, and grasped a water pipe to pull himself along. All at once the 1820 kHz signal disappeared. He released the pipe and the signal reappeared. Moving slowly along the pipe, the investigator discovered a gas pipe that crossed the water pipe, gently touching it. Moving the water pipe broke the contact between the pipes which was forming a nonlinear rectifier joint. The investigator placed a piece of plastic tape between the pipes, crawled out, dusted himself off, and received the congratulations of the excited radio amateur whose visions of winning the radio contest danced in his head.

Upon further thought, the investigator concluded that mixing occured between two strong broadcast signals, one on 1080 kHz and the other on 740 kHz. Their sum product (1080 + 740) equalled the interference frequency of 1820 kHz. The investigator also knew that other mixes of interfering signals could occur, particularly if the signals themselves had harmonic content (Figure 2).

The Case of the Musical Trash Chute

The RFI investigator solved a very unusual case of cross modulation and mixing that irritated a large number of people. He was called by an agitated apartment owner who reported that all of a sudden the radios in the large complex were rendered useless by a jumble of music and voices that was caused by some nearby radio ham or CBer. Could this individual be located and put off the air?

The investigator evaded the pointed question but promised he would look after the radio racket. Upon speaking to some apartment dwellers, he found that whenever the door of the trash chute to the apartment was opened, discordant sounds, music and voices could be heard! Possibly they came from the basement area where the trash was stored or collected?

The investigator carefully examined several of the trash chutes. He could actually hear the voices and the music. After an exhaustive investigation he determined that a loose strap holding a chute to the metal wall of the building was rectifying the signals from three nearby broadcast stations. Further, the metal chute was acting as an antenna, reradiating the signals which were then coupled into the internal wiring of the building, and thence into the radios in the building. And finally, the metal sides of the chute itself were acting as a loudspeaker!

Firmly fastening all straps holding the trash chutes to the frame of the building cured the RFI problem. The investigator told the apartment manager that metal drain spouts, metal lath, chicken wire in stucco walls, loose ground clamps and even metal roofs have been known to cause cross modulation and to "sing" in a strong rf field. Since the apartment was in a strong signal area, the manager would have to be continually on the lookout for inadvertent cross modulation and mixing in the building and its environs. Case closed!

Self-inflicted Television Interference

Today's television receiver is a complex device having multiple signal and control channels, plus various oscillators and sweep generators all running in a compact, unshielded cabinet, mass produced for the cutthroat consumer market. Because the tv set is a design compromise with cost an important factor, unwanted spurious signals generated in the tv receiver often cause interference to the picture and can cause interference to a nearby tv set, stereo or radio. That the tv receiver may have built-in interference problems is not known to the purchaser, who is quick to blame reception problems on a nearby radio amateur or CB operator. The following sections discuss some of the more common causes of self-inflicted TVI.

Co-channel TVI

There are only a limited number of vhf tv channels available and the same channel may be assigned to two or more stations. To reduce interference between stations on the same channel, assignments are usually limited to stations 150 miles apart, or more. Normally this physical separation is sufficient to prevent interference because signals in the tv spectrum normally do not follow the curvature of the earth.

Under certain atmospheric conditions, however, it is possible for tv signals to travel a considerable distance and "ghost" signals will appear under the local channel, creating annoying picture interference. This interference usually occurs on the lower channels, particularly channel 2 (Figure 3). The interference creates a fuzzy image because the tv channels are not on the same exact frequency and the frequency difference is seen on the screen as horizontal bars, moving vertically up the screen. This is known as the "venetian blind" effect.

In some areas of the country where co-channel RFI is prevalent, the interfering stations are synchronized in frequency to eliminate the bars, or one picture transmitter is offset from the other by 10.5 kHz to make the picture bars less objectionable.

When atmospheric conditions create co-channel TVI there is little that can be done except to try a more directive TV antenna which may help to cut down the signal from the offending station.

Fig. 3 Co-channel RFI is caused by "ghost" signal from distant tv transmitter appearing with local channel picture. Because the two stations are not on exactly the same frequency an interference pattern known as the "venetian blind" effect is created.

Tv Cross Modulation

A particularly exasperating form of internal cross modulation occurs in some tv sets and is caused by two or more strong tv signals entering the set and being mixed in a nonlinear stage of the receiver. Because of the nature of the television video signal, the receiver must be able to accept a very wide transmission band of frequencies for any one channel. Because it can do this, the receiver is susceptible to signals in other channels which may cause problems within the receiver. To a greater or lesser degree, adjacent channel signals are rejected but the picture information on one channel may be transferred by cross modulation to another channel, as shown in Figure 4. When the interfering signal is weaker than the wanted signal, a black bar usually appears vertically on the right side of the picture, sweeping to the left side of the picture (Figure 5). This is termed the "windshield wiper effect".

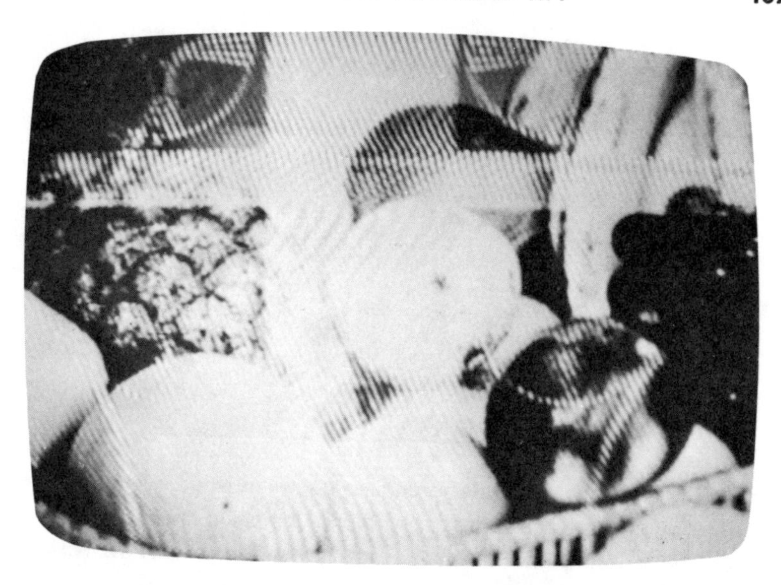

Fig. 4 Cross modulation effect from adjacent-channel strong signal results in unwanted picture seen along with desired channel picture.

Fig. 5 When interfering signal is weaker than wanted signal, black bar may appear on right of picture, sweeping slowly to the left.

Fig. 6 Adjacent channel interference is common when signals from two different cities are received near midpoint between them. New Jersey viewers receive this interference from transmitters in New York and Philadelphia.

Adjacent Channel Interference

Because of the problem of adjacent channel interference, tv stations are not assigned adjacent channels in the same area. But even though there is a good separation between channels in use in one area, adjacent channel interference can be noted in fringe areas, or when atmospheric conditions permit reception of tv signals from a distance (Figure 6). For example, channel 3 is assigned to Philadelphia and channel 4 to New York City, which are 90 miles apart. Viewers in New Jersey, midway between the two cities, often experience adjacent channel interference. Modern tv receivers are quite good at rejecting this form of interference but in some cases the unwanted signal is so strong it cannot be completely suppressed by the receiver. Antenna alignment can often cure this problem.

Fm Interference

Fm image interference does not affect a modern television receiver having a 45.75 MHz if system, but harmonics of an fm signal, either generated accidentally in the transmitter, or by a nonlinear device near, or in, the television receiver can create severe RFI to channels 7 through 13. The pattern produced is a "herringbone weave" on the picture, as shown in Figure 7. The pattern varies with the sound on the fm signal,

Fig. 7 "Herringbone weave" pattern on tv screen is caused by interference from nearby fm broadcast station. This interference is limited to older sets and may be cured by addition of fm trap filter in the antenna circuit.

and not with the sound or picture transmitted on the tv channel. Fm interference can usually be tuned out with an fm trap at the television receiver antenna input terminals. Some of the available brand-name traps that can be purchased through a large television dealer are: Blonder-Tongue, Winegard, Drake and Jerrold.

Local Oscillator RFI

The fm receiver can cause picture interference in a nearby tv receiver (Figure 8). Modern fm receivers tune the range of 88--108 MHz and the local oscillator operates 10.7 MHz above or below the signal. The oscillator tuning range, therefore, is either 98.7--118.7 MHz or 77.3--97.3 MHz. The oscillator also generates second harmonic energy in its particular tuning range. Oscillator radiation can thus fall in tv picture channels 5 and 6 or channel 12, causing the interference pattern shown in the photograph. The interference will come and go as the fm listener changes stations.

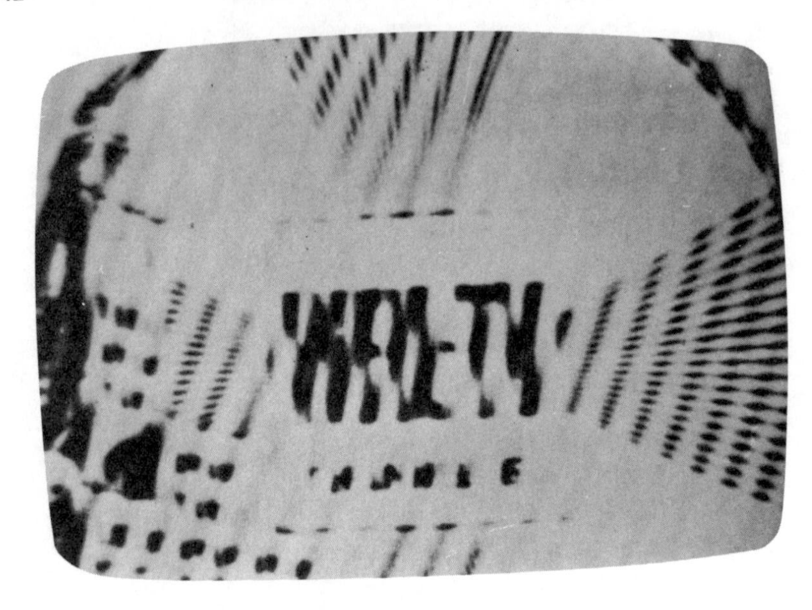

Fig. 8 Picture interference caused by local oscillator of a nearby fm receiver. Interference comes and goes as listener changes stations. Fm trap or antenna stub on the fm receiver will help cure the problem.

To eliminate this source of interference the serviceman will place an antenna stub on the tv receiver that rejects the spurious signal from the fm receiver oscillator. Separating the fm antenna from the television antenna is also helpful.

Barkhausen TVI

Some old television receivers can cause interference to themselves. One cause of such a problem is a *Barkhausen* oscillation (named after the German discoverer of this unusual phenomenon). The picture of such an oscillation is shown in Figure 9. This vertical black line (or lines) appears on the left edge of the screen and is caused by bursts of radio energy which fall in the rf band of the picture. The picture interference is created by the horizontal sweep output tube in older tv receivers. Under certain operating conditions, the tube breaks into pulsed oscillations which are a function of the electrode voltages of the tube. The oscillations are picked up in the rf section of the receiver as a spurious signal and can be

Fig. 9 Old tv set causes interference to itself. Vertical black line is caused by oscillation in horizontal sweep tube. Adjustment of horizontal drive control or replacement of tube cures the problem.

eliminated by adjustment of the horizontal drive control, or by replacement of the sweep tube. These oscillations sometimes occur when a "rabbit ears" antenna is used as it is commonly placed atop the receiver, near the sweep circuits, causing a feedback path for the oscillator circuit. Moving the antenna away from the receiver sometimes cures the problem.

The Tv Snivet

Have you ever seen a *snivet* on your tv screen, as shown in Figure 10? This is a spurious signal created by sudden changes in the current drawn by the horizontal sweep output tube. Replacement of the tube eliminates the snivet.

Horizontal Oscillator RFI

A common source of interference that can cover a wide area around a television receiver is horizontal oscillator radiation. The oscillator

Fig. 10 Spurious "snivet" on tv screen is eliminated by replacing the horizontal sweep output tube used in older tv receivers.

generates watts of power on 15.75 kHz and the sweep signal is rich in harmonics which are heard as rough signals every 15.75 kHz across the dial of a nearby broadcast or shortwave receiver. To make matters worse, the harmonic signals can be radiated by the tv antenna and lead-in wires and cause problems in nearby receivers. Sooner or later, an interference complaint traceable to a television receiver oscillator will be logged by the local power company or the FCC.

The RFI investigator received a written complaint of radio interference from a renter in a large apartment complex. He called the complainant and listened to the noise over the telephone, immediately identifying it as sweep oscillator interference. The investigator told the complainant that it was his responsibility to locate the source of interference and that he could do it with a portable radio. The complainant replied that there were over 200 apartments in the complex and each one of them had one or more television receivers.

The investigator expressed his sympathy and suggested that the complainant drive around the complex and listen to the noise on his car radio to see if he could determine the area in which the interference was the loudest.

A few days later the complainant called to tell the investigator that he had located the noisy tv set only two apartments away. The investigator told him to buy a high pass tv filter and have the owner install it on the back of the offending tv set where the antenna connections were made. This would reduce the level of spurious signals travelling up the antenna and radiating to nearby receivers. A later phone call confirmed that this was done and the interference had dropped to a very low level.

Color Burst Interference

Modern color television receivers have an internal oscillator that is frequency-locked to the master oscillator of the tv transmitter. Frequency lock is accomplished by means of special signal transmitted between picture signals. The frequency of the local oscillator in the receiver is approximately 3579.5 kHz. This signal, and its harmonics, fall in the radio amateur bands and can often be heard miles away from the offending tv set. Multiple signals can be heard from local television receivers and the signals mix among themselves and cause perplexing interference. Again, a high pass tv filter on the set will help prevent these oscillations from being radiated by the tv antenna and lead-in.

In some cases the color burst oscillator can go into an unstable, parasitic oscillation that will cast vertical bars extending midway across the screen, plus bands of shot lines on the screen of the defective set. The bars are more noticeable when a color set is receiving a black and white transmission and less so when a color picture is received. This spurious emission can be corrected by a serviceman, usually by replacing the oscillating crystal in the receiver. Harmonics from the color burst oscillator can often cause interference to a nearby tv receiver up through channel 13.

Tv "Ghosts"

Television signals, like radar signals, can be reflected from large surfaces. In some instances a clear picture can be distorted by a ghost that will appear overnight. The resulting picture resembles a duplicate picture displaced a fraction of an inch to the left or right of the original picture (Figure 11). The ghost signal may be caused by reflection from a nearby building, power line or hill. An "instant ghost" that seems to show up suddenly may be created by the erection of a new television, ham radio or CB antenna in the vicinity of the tv antenna. Reorientation of

Fig. 11 "Ghost" picture on tv screen is caused by signal reflection from nearby hills or buildings. Reorientation of the tv receiving antenna will help to eliminate the unwanted "ghost" signal.

the tv receiving antenna, or moving it a few feet on one direction or another will help to eliminate such a ghost signal.

Channel 6 Chroma Beat

An interference pattern can be observed on a color tv set caused by the channel 6 video carrier (83.25 MHz) and the if sound carrier (41.25 MHz) beating with the if chroma (color) carrier at 42.17 MHz (Figure 12). The interference pattern drops out if the color intensity control is turned down. A television technician can make adjustments to the receiver rf delay circuits to eliminate this source of interference.

Channel 8 Interference Pattern

This picture interference is developed in the detector circuits of the tv set when the fourth if harmonic of 45.75 MHz at 183 MHz produces a beat located in the center of the channel 8 response curve. A swirling, "S" pattern is produced that locks in position when the automatic frequency

Fig. 12 Color interference drops out if color intensity control is turned down. Tv technician can adjust receiver to eliminate this fault.

Fig. 13 Tv picture interference on channel 8 can be cured by moving tv lead-in wires away from the set or removing "rabbit ears" antenna from the top of the receiver.

Fig. 14 The interference pattern created by internal circuits of the receiver. Elimination requires replacement of defective component in the retrace circuit by a competent serviceman.

control (AFC) is activated, and drifts about when the AFC is off (Figure 13). The interference is mainly brought about if the antenna lead-in passes near the picture detector in the receiver. Moving the lead-in wires away from the set clears the picture.

High Frequency Transistor Radiation

A vertical line, resembling a snivet, appears on the tv screen, moving back and forth with the passage of time (Figure 15). The interference is usually produced in one set and transmitted to others. Usually it is not visible on the set generating the condition. This oscillation takes place in the sync separator or agc gate transistor and placing a finger on the case of either low voltage transistor reduces the oscillation. The problem can be eliminated by placing a ferrite bead on each transistor lead or installing a 100 pF capacitor from collector to base.

Zener Noise Interference

The Zener noise resembles fine snow on the picture (Figure 16) and is caused by random noise generated by one or more diodes in the receiver. A .01 uF capacitor placed across the Zener diode will cure the problem.

Fig. 15 Transistor oscillation in tv receiver produces vertical line on the screen. Problem can be eliminated by circuit modification performed by a serviceman.

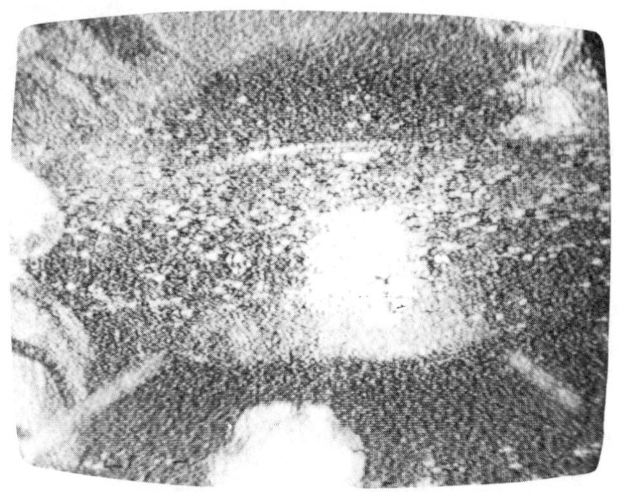

Fig. 16 Zener diode noise interference resembles fine snow on the tv screen and requires addition of a capacitor by a serviceman.

Cure Your Own Self-inflicted TVI?

In some cases the problems discussed above may be related to receiver alignment or may be peculiar to a certain brand or model of receiver. In many cases, suggested "fixes" are known by the service technician or manufacturer of the receiver. It is always a good idea to make certain that there is no malfunctioning circuit in the receiver itself before a nearby amateur or CB operator is blamed for the problem.

A quick check for receiver operation can be made by simply sliding a hand along the lead-in while observing the effect on the picture. If the picture changes considerably as the hand is moved, the antenna installation should be checked before any modifications are made to the receiver.

Interference can also be caused if the agc circuits are functioning improperly; for example should the if bias be too low and the rf bias too high, the picture becomes excessively snowy and may be subject to rf beat interference.

Alignment of the receiver must be correct if interference is to be minimized. The sound trap and adjacent channel traps must be properly adjusted.

Because of the complexity of the problem it is recommended that RFI which seems to be self-inflicted be left to a competent service technician who knows the receiver and is conversant with the idiosyncracies of the particular set.

Interference can be created by sources that are not normally suspect, even by a trained RFI investigator. The RFI can be created by some everyday devices that may suprise you.

Cruise Control RFI

In some late model automobiles the cruise control feature is affected by a strong rf field causing a sudden increase or decrease in speed of the vehicle. The effect has been noticed in some Cadillac and Mercedes-Benz automobiles. In these cases the electronic fuel injection system and skid control system may also be affected by a nearby mobile transmitter. Manufacturers of the vehicles are aware of these problems and modification kits are available to eliminate these suprising and potentially dangerous effects of RFI.

(Note: Cadillac owners with cruise control that disengages with RFI should request a "Suppressed Speed Sensor", part number 646-6900 from their dealer. It is a no-charge modification).

Fig. 17 Light bulb RFI! Certain 25 watt bulbs can produce strong interference to tv channel 2. Offending bulbs should be returned to place of purchase for a refund or replacement.

Light Bulb RFI

Did you know that everyday, ordinary lightbulbs can create RFI? They certainly can, and do!

Very early bulbs having a clear glass envelope and straight wire filaments and a tip on the end of the bulb are prolific RFI generators, producing a rough, unstable signal in the 60 to 70 MHz range. This signal is full of harmonics and can be identified as a horizontal black line across the tv screen (Figure 17) Because of the length of the filament, one end of it can be positive when the other end is negative and electrons can oscillate back and forth as the polarity of the ac line reverses itself. A high frequency oscillation is set up, modulated at a 120 Hz rate.

Even though production of these bulbs ceased about 1925, they still can occasionally be found in cellars, closets and attics of older homes, When one is found, it should be removed and carefully saved as it is an antique worthy of collection!

Recently, the investigator has traced RFI problems to modern lamp bulbs. The problem appears as two horizontal bars or lines on the tv screen, approximately two to three inches apart. The bars appear only on channel 2, slowly moving vertically up the screen. When the bars disappear at the top, two more bars appear at the bottom of the screen.

An exhaustive examination pointed to a nearby lamp as the culprit.

Different bulbs were tried in the lamp and it was found that the RFI producing bulb was a 25 watt unit (of a popular brand). Higher wattage bulbs did not cause the problem. Out of a carton of four 25 watt bulbs, only one was found that did not cause RFI on TV channel 2.

Lamp engineers have been notified of this problem and they realize the condition exists but, to date, nothing has been done to the design of the bulb to eliminate this source of RFI. The reader is warned of this problem and when the condition is found, it is suggested that the offending bulbs be returned to the place of purchase for a refund. Be sure to mention that the bulbs are a source of television interference.

RFI From A Master Clock

Have you ever noticed the wall clocks mounted in schools, factories, offices, hospitals and municipal buildings? The investigator has and noted that the sweep second hand normally pauses for several seconds, or might jump for 10 to 15 seconds in one second intervals. Such clocks are "slaves" of a master clock which sends a 500 Hz timing pulse through the wired clock system to hold all clocks to the correct time. This 500 Hz tone, unless carefully isolated by networks and transformers, can get into the building lighting system and be conducted to nearby tape recorders or video recorders. This will disrupt the recording and, unless the victim of the interference is aware of the source, can cause mysterious problems on tapes and tape decks. The solution is to contact the suppliers of the clocks who can install suitable filters and suppression circuits to limit the interference.

The Case of the Whistling Signal

Two calls came to the attention of the RFI investigator one day. Both of them were from radio amateurs. One operator complained about a weak whistling signal on the 20 meter band heard at 14.318 MHz during the evening hours. The other amateur heard the same sound on the 40 meter band at 7.159 MHz. The two amateurs lived about two miles apart and both were experiencing the same form of interference on different amateur bands.

The first amateur, using his beam antenna, thought the RFI was west of him. The second amateur reported the interference very loud but could get no directional peak on it as he had no beam antenna. From this scant information the investigator deduced the source of RFI must be closer to the second amateur.

The investigator asked both amateurs to check 3.579 MHz, the fundamental of both noise frequencies--and also the frequency of the color burst oscillator in a television set. The investigator told both amateurs he thought the noise they were hearing was the color burst oscillator feeding back into a television antenna and radiating into the surrounding area. The amateurs told him that they thought they could locate the source of interference themselves.

Sure enough, one of the amateurs called back a few days later and told the investigator that he had found a defective tv receiver a few houses away from the 40 meter ham operator. The owner admitted he was having color problems with his set and had already called a serviceman to look it over. The amateurs suggested that he install a high pass filter on his receiver for good measure once the set had been serviced. The owner promised to do this and the problem was solved by a few phone calls by some observant and friendly radio amateurs.

The Case of the Roaring Signal

The RFI investigator was notified of a complaint of a "roaring noise" heard by a radio amateur operator. The noise started about 3 pm and continued into the evening hours. A similar complaint was received from a woman listener in the area who complained that the evening news was blanked out in her broadcast receiver by "rumbles".

The investigator called the radio amateur and asked him to patch the interference to him over the telephone. As soon as he heard the sound, the investigator recognized it as being created by the horizontal sweep oscillator of a television receiver. He told the amateur that the sweep oscillator operated on 15.75 kHz and the output was very rich in harmonics that could be heard every 15.75 kHz across the spectrum in the vicinity of some television receivers. The rough signal sounds like a roar or hiss that overrides weak stations in the broadcast band. The amateur was affected much more because of the low-level signals he was trying to pick up.

The investigator sympathized with the radio amateur but told him it was not the responsibility of the power company and suggested that the amateur try and locate the source of interference by driving around his neighborhood and listening to the car radio. He might be able to determine where the interference was loudest.

A week later the amateur called the investigator and told him he had located the tv set causing the problem. He had used a portable radio and had pinpointed the source, which came on a few moments after a neighborhood woman returned from work.

Following the investigator's advice, the amateur called the woman on the phone and found out that she had recently complained to the power company about a "rumble" in her kitchen radio. The amateur suggested she contact her tv serviceman and have the set checked and have a high pass filter installed at the antenna terminals on the back of the set. He explained that the high pass filter worked two ways. It prevented his ham signal from entering the tv set and it helped prevent the sweep oscillator radiation from the set from jamming her kitchen radio and also his reception.

A few days later the amateur noticed that the sweep oscillator noise had been greatly reduced at his receiver and concluded that the tv set had been serviced.

The investigator remarked that interference from tv sweep oscillators was a growing and serious problem. The noise is radiated from the tv antenna and lead-in and also is coupled back into the power line by induction from the antenna field and also by conduction down the power cord. He noted the interference level was highest in homes that had "knob-and-tube" wiring and less in homes that were wired in metal conduit. He also told the amateur that a line filter placed on the power line to the tv receiver often helps this serious problem. But he emphasized that the problem could not really be cured until the receiver manufacturers made an effort to contain the powerful harmonics of the sweep oscillator. He didn't see this happening in the near future.

The Silicon Controlled Rectifier

The silicon controlled rectifier (SCR) is a semiconductor that, in its normal state, blocks a voltage applied in either polarity or direction (Figure 18). The device conducts in a forward direction when a control signal is applied to the gate electrode. When conduction is established it continues after the control signal is removed until the anode supply is removed, reduced or reversed. SCRs by the millions are used for lamp dimmers and speed controls for electric tools and a myriad of small powered devices.

Because the SCR is continually controlling the line voltage by switching from a very high impedance to a low one on each cycle in about a microsecond, it produces a waveform rich harmonics, creating severe RFI up to about 7 MHz. The characteristic RFI noise is loud buzzing that completely covers the lower frequencies, but quickly drops in intensity in

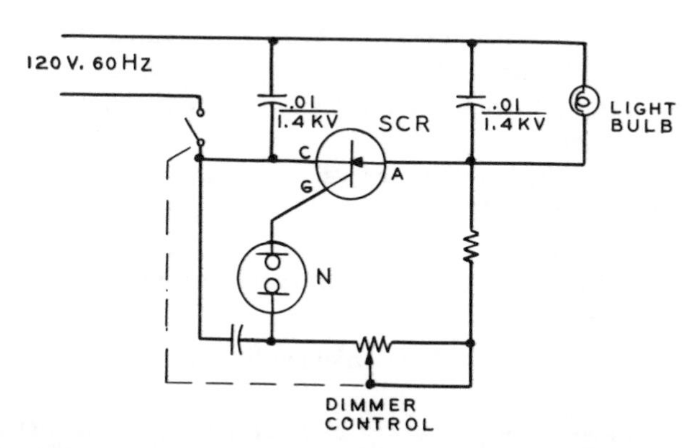

Fig. 18 The silicon controlled rectifier (SCR) in a light dimmer is a prolific source of RFI. New dimmers in metal case have internal filtering to reduce radio noise. Older dimmers can be RFI-suppressed by addition of .01uF, 1.4kV disc ceramic capacitors across input and output terminals.

the vicinity of the 40 meter ham band. Dimmers can be purchased that have built-in RFI filters but they are more expensive than filterless dimmers and consequently are not as popular. The better dimmers are built in a metal case, so beware of a dimmer in a plastic case--chances are that it has no RFI filter circuits.

Those plagued with a noisy dimmer control can either replace it with an interference-free design or can alleviate the problem to a degree by bypassing all external leads to the ground wire by means of .01 uF, 1.4 kV working voltage ceramic disc capacitors.

The Case of the Heliarc Welder

A radio ham called the RFI investigator one day and complained of interference on the 40 and 80 meter bands that resembled a loud hissing, or frying sound. The investigator visited the ham the next day and heard the unusual noise on his portable receiver and suggested that it probably was a Heliarc welder. He also told the ham that he had passed a shop about five blocks distant that fabricated aluminum parts for aircraft; he suggested the ham look there, as the problem was not within the jurisdiction of the power company.

The ham visited the shop and introduced himself to the foreman. He was shown several welders, one of which had the metal sides removed from the cabinet. Bringing a portable receiver near the welding machine produced a heavy blast of radio noise. The ham told the foreman that removing the covers had taken away the shielding that restricted the radiation of radio noise. The foreman agreed to replace the side shields and also agreed to connect the frame of the machine to an external ground (see Chapter 11).

Later, when the investigator was in the area, he found that the interference from the welder was no longer in evidence. He knew that the Heliarc welder properly installed and maintained would not cause appreciable RFI. Now the radio ham and foreman knew this, too.

The Shielded Room

Often times it is necessary to have leads of considerable length connecting the work load to the rf generator or welder. These leads radiate rf and so shielding the noise source and filtering the power line are not enough. The entire work area should be shielded under these conditions. Other installations requiring the complete shielding of the work area are induction heaters fed by conveyor belts. Here the belt should enter and leave the shielded area via a shielded tunnel.

The most economical solution to the RFI problem in such cases is the installation of a shielded *screen room.* Such rooms are commercially available from several manufacturers. The screen rooms are manufactured in modular segments that fasten together to provide rooms of various sizes. A screen room of the "cell" type usually has a guaranteed attenuation of 100 dB from 14 kHz to 1,000 MHz. Filters can be supplied which isolate power lines and control wires entering or leaving the room.

Chapter 9

What You Can Do To Your Transmitter To Lessen RFI and TVI

Fifty years ago the only home entertainment devices were the windup phonograph, the battery radio and the stereoptican slide viewer. The modern home, on the other hand, is often a vast intermix of electronic devices of awesome complexity: telephones, am/fm radios, television receivers, fire and smoke detectors, intercoms, electronic organs, stereo equipment, tape decks, hearing aids, video recorders, small computers, garage door openers, radio amateur and CB equipment and so on. And all of these modern miracles can be affected by RFI. The sources of RFI are plentiful, too, but largely unknown to the public. The early chapters of this handbook list many of the sources.

Sometimes the ham or CB operator is at fault, creating objectionable RFI. Some CBers use illegal high power amplifiers to boost their signals and create severe neighborhood interference because of the poor design of these cheap devices. Hams and CBers have been known to have faulty or improperly adjusted equipment. But, more often than not, the RFI problem lies with the home entertainment equipment due to poor installation, insufficient shielding and filtering and other cost-cutting manufacturing techniques.

To make matters worse, the proud owner of a new electronic gadget resists the suggestion that his equipment is at fault and is quick to point out that he wouldn't have interference if it wasn't for that ham or CBer down the block who is disrupting his enjoyment. Since his gadget functions properly when the ham or CBer is off the air, it obviously isn't the fault of his gadget. Right? Wrong.

A Difficult Confrontation

Without entering into the philosophical implications of RFI problems, it is extremely difficult for the user of home entertainment devices to concede that his expensive equipment may be at fault, and it is equally difficult for the amateur or CBer to convince his neighbor that merely because he spent eight hundred dollars for his stereo, and somebody down the street has a transmitter, the stereo is acting as a shortwave receiver.

The unhappy owner resists the idea that, while modern tv and stereo equipment is not intentionally designed to pick up nearby radio signals, it can do just that because of economic considerations (in a highly competitive market) that discourage protective circuits which reject undesired signals. The cost of such circuitry would increase the cost of the device a bit, and besides, less than ten percent of the total number of home entertainment equipments in use today suffer from RFI--or so the manufacturer's story goes.

This is cold comfort to the viewer or listener bedeviled by RFI but most manufacturers believe it is unfair and unnecessary to burden the mass consumer market with the extra cost of RFI-proofing. On the other hand, some manufacturers, dealers and service men have devised procedures to improve the RFI-rejection of their home entertainment devices. Unfortunately, this information is not widely distributed or fully recognized. Nor is it known by one in a thousand owners.

Neither is it fully realized that an entertainment device may perform the task it was designed to do but still require special treatment to reject strong, undesired radio signals.

The Ever-changing RFI Problem

The RFI situation at any given location is not necessarily stable. For example, a user may live in an area free of amateur and CB operators and far enough from local broadcast and TV stations so that his equipment has no difficulty in rejecting weak unwanted signals. The situation may change overnight if a ham or CB operator moves into the neighborhood, or if a new industrial, police or other service radio station is activated in the vicinity. The equipment, which required no special treatment before, now requires modification to reject the new signals. Or, on the other hand, the user may move into an area where strong radio signals are present and the equipment that was satisfactory in his old location is now prone to RFI. Either situation creates a problem between the user of the entertainment device and the nearby radio station.

In the main, RFI problems are solvable, providing sufficient time, expertise and money are brought to bear on the situation. Fortunately, most RFI cases are relatively easy to reduce. But even if the interference is the fault of the home entertainment device, the amateur or CB operator is usually judged guilty until proven innocent and may unwittingly becomes the victim in a nasty neighborhood dispute unless he is wise enough to protect himself *before* the trouble arises.

What Can the Ham or CBer Do?

The reaction of a ham or CBer to an RFI complaint will determine his relationship with his neighbors for months to come. Thus it is very important that a head-on confrontation be avoided and the situation handled in a cool, cooperative, low-key manner.

The first and most important consideration for the ham or CBer (even *before* he has an RFI complaint) is to make sure his transmitter is operating properly and free of spurious rf products, including TVI-producing harmonics. Don't wait for the complaint of an angry neighbor! It is imperative that the television receiver and other entertainment devices in the ham or CBer's home be free of RFI problems because it is not very convincing to claim that the interference is due to your neighbor's equipment when your own transmitter is not as "clean as a hound's tooth". On the other hand, it is a powerful argument to show that your transmitter does not cause any RFI problems to your own home entertainment equipment. So if you have RFI problems in your own home, it is important to eliminate them before you take any other steps, or before you are accused of RFI by an unhappy neighbor.

Once your transmitting equipment is RFI-free, you can then demonstrate that you have no RFI problems and it will be easier to convince your neighbor that his equipment--not yours--is at fault. In any case, the burden of proof falls on the ham or CBer to make sure his equipment is clean and meets the FCC standards concerning spurious radiations.

In general, modern transmitting equipment commercially manufactured and type-accepted by the FCC has circuits and shielding built into it to reduce spurious emissions. Even so, you should insure that your own equipment is working properly and check it out for RFI.

As with home entertainment equipment the fact that your transmitting equipment may have cost upwards of many hundreds or even thousands of dollars is no assurance that it is TVI or RFI-free.

Amateurs or CBers living in a fringe tv area are confronted with a

BAND	TV CHANNEL				
	2	3	4	5	6
20 M	4TH	—	5TH	—	6TH
15 M	—	3RD	—	—	4TH
10 M	2D	—	—	—	3RD
CB	2D	—	—	3RD	3RD

Fig. 1 The relationship between ham bands, CB channels and tv channels. Harmonics of 20, 15 and 10 meter amateur bands plus the CB channels fall into tv channels 2 through 6. Radiation of transmitter harmonic energy can cause severe TVI on these tv channels.

difficult problem. Even the best TVI reduction measures may not clean up an interference problem to the satisfaction of the neighbors. Unfortunately, some transmitting equipment is more prone to RFI than others and your last resort may be to sell the offending equipment and get other gear that has less of a harmonic radiation problem! Since specific information about harmonic radiation of most transmitters is difficult to obtain, this can be a hit-or-miss solution that could cost much time, money and effort.

Low power (QRP) amateur equipment is a solution for many hams who live in apartment houses or other difficult RFI areas. These rigs cause almost no TVI even under difficult conditions and get out well when the operator possesses patience and operating know-how. The higher the transmitter power, the greater the RFI problems.

How to Check Your Transmitter

The first thing to do is to place a color tv set in the same room as your transmitter. Connect it to a small antenna such as "rabbit ears" or an equivalent antenna that will provide a reasonable, yet weaker, tv signal than your neighbors would have. Check normal operation of your transmitter to see if you cause TVI.

Now connect the transmitter to a dummy load with a short length of coaxial line. Make sure you use the appropriate connectors on each end of

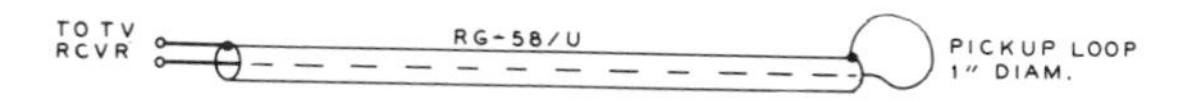

Fig. 2 "Sniffer loop" attached to your tv set shows harmonic leakage from your transmitter. Pass the loop around the transmitter cabinet and watch tv screen for indication of RFI.

the line so that line leakage will not interfere with your observations. If there is no interference on the harmonically related channels (Figure 1) after you have made the following test, the shielding of the transmitter is adequate. If interference is experienced on a harmonically related channel, the transmitter shielding is inadequate.

The instrument used to check harmonic radiation is a "sniffer loop" made up of a short length of RG-58/U coaxial cable with a single turn loop about one inch in diameter soldered between the center conductor and the shield at one end (Figure 2). The other end of the line is connected to the antenna terminals of the tv receiver (receiver antenna removed, of course).

Energize your transmitter on your operating band, consult the chart of Figure 1 and observe the various tv channels listed for the harmonics that may cause interference on the channels. Run the loop around the transmitter case observing where radiation is strong, such as ventilation slots, meter and dial openings, line cord and power supply leads. Checks should be run on the 20, 15 and 10 meter bands first. The lower bands (160, 80 and 40 meters) also have harmonics which fall in various tv channels but they are much weaker, so for the purposes of checking transmitter harmonic interference, it is only necessary to consider the higher bands. The CB operator should test his equipment on all channels.

If you have no TVI on the "rabbit ears" antenna and can detect no harmonic pickup with the loop, your transmitter is free of objectionable harmonics. On the other hand, if you detect harmonic energy it may be possible to reduce it by minor modifications to your transmitter.

Cleaning Up the Transmitter

Once you have made the checks, list any interference detected and on what tv channels and transmitter operating frequencies it occurs. Next,

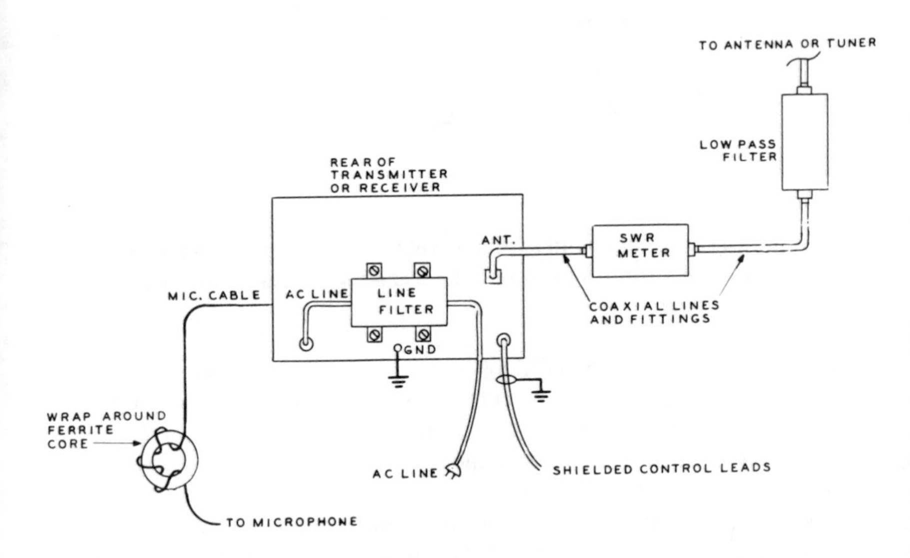

Fig. 3 Proper filtering of cables from the transmitter will go a long way in the battle of cleaning up TVI. Low pass filter should be placed after SWR meter and before any antenna tuner.

operate your equipment normally into your antenna on all bands you use. Have a family member or friend check the entertainment devices in your home for interference. Make this test when the band is "dead" and make sure that you are not interfering with other operators. Make the tests short.

When you have completed these tests you will have a good idea of the harmonic content of your transmitter and which tv channels are affected by it.

The first step in cleaning up your transmitter is to place a *low-pass filter* in the coaxial antenna cable of the transmitter and to add a *line filter* in the power cord to the ac outlet. Data on a line filter is given in Chapter Two. The case of the line filter should be mounted to the case of the transmitter or power supply and grounded to it.

The low-pass filter is inserted in the coaxial cable after any SWR meter, as the diodes in the meter can be a source of harmonic radiation (Figure 3). If you use a linear amplifier, it is a good idea to use two low-pass filters, one between the exciter and the linear and one between the linear and the antenna.

You should now remove your antenna and check operation of the low-pass filter using a dummy load. If you are clean on the dummy load and not interference-free on your antenna, the filter may not be doing its job. This could be because harmonic signals are flowing on the outside shield of the coaxial line and over the outside case of the filter.

The next task is to reduce harmonic currents that occur on the outside of the transmitter and transmission line. The rf "sniffer" will give you an idea of where the unwanted harmonic radiation is coming from. The low-pass filter and line filter "bottle up" the harmonic energy in the transmitter and it now radiates from the openings and connections to the equipment.

If the transmitter has an external power supply, the multiple conductor cable should be shielded by pulling flexible, shielded braid over the leads and grounding the braid to the transmitter chassis and power supply chassis at each end of the cable.

If an external speaker is used with a transceiver, the cable should be short and made up of a shielded wire, the shield serving as one conductor and grounded to the transceiver chassis and the speaker frame. Bypass the "hot" speaker lead at the voice coil with a .01uf ceramic capacitor. An additional capacitor may be required across the terminals at the transceiver.

Keying and control leads should be run in shielded wires with the shield grounded firmly at the equipment. It is not advisable to bypass these leads as the capacitors may interfere with equipment operation.

The microphone cable should be as short as comfortable, an eight-foot cable being a quarter-wavelength at 10 meters. Two or three very small ferrite beads (Amidon FB-43-101) slipped on each microphone lead inside the transmitter case may help to cool a "hot" microphone. Alternatively, two or three turns of the microphone cable may be wrapped around a ferrite core (Amidon FT-82-43) to break up any incidental radiation from the line.

The power cord (in addition to having a line filter in series with it) can act as an antenna, picking up rf energy radiated by your station antenna and feeding it directly into the nearby tv set via the house wiring. Wind the power cord tightly around a ferrite rod (Amidon R61-050-750) and tape it in place.

If an external antenna relay is used, the relay coil should be bypassed with a .01uf, 1.4kV ceramic capacitor and the relay power cable shielded. The shielded wires should be wrapped tightly around a ferrite rod and taped into position. (Note that the shielded wires should have outer insulation or the turns will short themselves out when wrapped on the rod).

In some instances the addition of an external antenna tuner (or "transmatch") will reduce RFI as it provides additional rejection of transmitter harmonics. The metal case of the tuner should be grounded to the equipment with a short, heavy lead. All of these modifications are shown in Figure 3.

The Transmitter Cabinet

This is the weakest link in the TVI-reduction chain of events. Some years ago the transmitter inclosure was a well-constructed rf shield. The old *Johnson Viking II* transmitter is an excellent example of this technique. Such expense is no longer incurred today, and modern transmitter enclosures are little more than dust shields. Trying to improve shielding on existing equipment is very difficult. The amount of shielding required depends upon the equipment design, the frequency involved, the strength of the tv signal and the distance between the equipment and the tv receiver.

At the very least, the lid of the case and the panel should make good electrical contact to the case if all are made of metal. A grounding clip on the front lip of the lid often helps. If the equipment case is plastic, you are out of luck.

Grounding the Transmitter?

Grounding the transmitter or the outer shield of the coaxial line to the antenna may or may not help in eliminating TVI. However, if you are ever visited by the FCC or a TVI-committee in response to a viewer's complaint, one of the first things they look for is a ground connection to the transmitter. No matter if it is a good ground or not, you will get a "black eye" if your equipment is ungrounded.

The problem with a ground connection is that the length of the ground lead from the transmitter to the point where the water pipe or ground rod actually enters the ground is a good fraction of a wavelength on the high frequency ham bands and the CB channels. If, for example, the ground wire is eight feet long, it is a quarter-wavelength on the 10 meter band and the transmitter is effectively insulated from ground just as if the ground wire was removed. If you can make a short ground connection (only a few feet) to a nearby cold water pipe, or a ground rod, it may help TVI, but don't be suprised if it makes no difference.

A gas pipe should never be used as a ground as insulated joints are used in the pipeline which provide no electrical continuity. Water pipes, especially copper ones, are better. But plastic water pipes are no good at

BAND	RADIAL GROUND WIRE LENGTH	
	FEET	METERS
160 LOW	123′ 0″	38.5
160 HIGH	120′ 0″	36.6
80	63′ 0″	19.3
40	32′ 6″	10.0
20	16′ 6″	5.3
15	11′ 0″	3.3
10	8′ 3″	2.0
6	4′ 6″	1.35

Chart of radial ground wire length for amateur bands. Radial length is nine feet for CB channels. Free end of radial wire should be taped to prevent shock hazard.

all. A quick check will tell you if your ground connection is useful. But useful or not, it is advisable to have it if your equipment is subject to inspection. See Chapter 11 for additional information.

The Radial Ground Wire

The radial ground wire is an artificial radio ground that can be effective. It can be used when the distance to a suitable ground point is too far. The radial ground is simply an insulated wire, one-quarter wavelength long at the operating frequency, connected to the transmitter ground terminal at one end and run away from the equipment in a random direction, either indoors or outdoors. The far end of the wire is left free. The end of the wire is taped to prevent contact as it is "hot" with rf energy and can cause a nasty rf burn if touched when the transmitter is on the air.

As a radial ground wire is resonant, it will only work properly on one band. Two or more radial wires may be attached to the transmitter for multi-band operation. Placement of the radial is not critical, although it is usually run in the horizontal plane, along the floor of the radio room tacked against the wall, or perhaps out the window and along the side of the building. For the lower bands where the radial is quite long, it can be run through bushes or trees a few feet above the ground. The radial should not actually touch the ground, nor any metallic object.

If a resonable ground connection is available, it should be used along with the radial. The combination of the best possible ground, plus a radial ground wire will go a long way in the reduction of television interference.

For multi-band operation, a multiple radial may be cut out of five conductor tv control cable, each wire being cut to the length of one radial for each band. Because of capacitance between the wires of the cable it may be necessary to trim the wires a bit shorter than the resonant lengths given in the table. In some instances a second set of radial wires will be of benefit in establishing rf ground at the transmitter.

Incorrect Transmitter Adjustment Can Cause RFI

Often overlooked as a cause of RFI and TVI is improper transmitter adjustment by the operator. A common error which causes TVI, or makes it worse, is overdriving the final amplifier stage in the transmitter. Reducing microphone gain or reducing drive to an amplifier may effect a dramatic reduction in TVI. Most transmitters and transceivers have a "drive" control or "microphone gain" control and dropping drive or gain to the proper level will not weaken the signal; on the contrary, it will make it more readable. "Power microphones" (with built-in amplifiers) are a source of overdrive and should not be used, or used with caution.

You can frequently reduce or eliminate RFI or TVI by reducing transmitter power in conjunction with the other recommended steps. Thoughtful amateurs and CBers in heavily populated areas often reduce power during the evening hours to reduce the chances of interfering with tv reception. High power is not necessary for enjoyable contacts as many amateurs and CB operators believe.

Sources of Ferrite Rods and Power Line Filters

As of this writing, the following manufacturers supply ferrite rods suitable for the construction of simple power line filters. Amidon Associates, 12033 Otsego St., North Hollywood, CA 91607. The 30-61-7 rod (7-1/2 inches long, 1/2-inch diameter) is recommended. The ferrite bead FB-73B-101 is suitable for use on microphone leads.

J.W. Miller Co., 19070 Reyes Ave., Box 5825, Compton, CA 90224 supplies a similar rod, FR-500-7.50. They also supply a miniature ferrite toroid core which may be used in place of a ferrite bead, part F-23-1.

Power line filters are available from:

J.W. Miller Co. (see address above). The C-509L filter is designed for insertion in the power line and can carry 5 amperes of line current. The C-508-L is a smaller version capable of carrying 3 amperes.

Other filter manufacturers are:

Cornell Dubilier, 150 Avenue L, Newark, NJ 07101.

Sprague Electric Co., 645 Marshall St., North Adams, MA 01247.

Shielded cables are available from:

Alpha Wire Corp., 711 Lidgerwood Ave., Elizabeth, NJ 07207

Belden Corp., Box 1100, Richmond, IN 47374

Write for the catalog or ask to see the catalog at your nearby radio parts distributor.

Sources of Low-Pass and High-pass Filters

As of this writing, the following manufacturers supply suitable filters:

J.W. Miller Co., 19070 Reyes Ave., Box 5825, Compton, CA 90224

Barker & Williamson, Inc., 10 Canal St., Bristol, PA 19007

R.L. Drake Co., 540 Richard St., Miamisburg, OH 45342

The William N. Nye Co., 1614 130th Ave. N.E., Bellevue, WA 98005

Radio Shack, Inc., Fort Worth, TX 76107. (Local stores carry filters or they may be ordered through the catalog).

Also--check with your local radio amateur or CB equipment dealer as many of them stock suitable filters.

Six Meter Harmonic Problems

The second harmonic of a six meter transmitter can cause interference to fm receivers and the third harmonic can cause interference to mobile services. A simple half-wave harmonic filter that can be easily constructed in the home workshop is shown in Figure 4. It provides about -35 dB rejection to the second harmonic and more to the third. For power levels up to 100 watts, silver mica style capacitors can be used. For higher power, the capacitors should be transmitting-type ceramic units of 5 kV rating.

The filter is built in a small metal or printed circuit board box with coaxial fittings on each end and a shield plate fastened across the interior. If the coil dimensions are followed carefully, no adjustment of the filter is necessary.

Two Meter Harmonic Problems

The second harmonic of a two meter transmitter falls into mobile communication channels and the third harmonic near tv channel 14. A simple tuned filter provides better than -40 dB rejection to all emissions outside the two meter band. The schematic and construction information

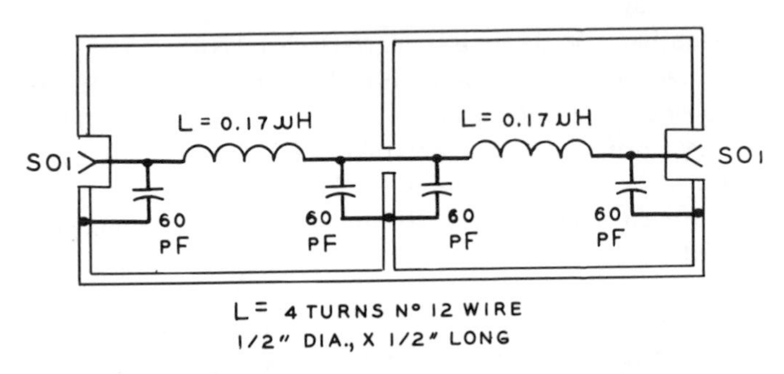

Fig. 4 Six meter TVI filter is built in two section shielded box which can be made up of double-sided printed circuit boards soldered together. The capacitors are dipped mica units for power levels less than 50 watts. Standard 62 pF or 56 pF values can be used in lieu of the specified 60 pF.

is given in Figure 5. The filter is built in a small cast aluminum box for maximum shielding, or one made up out of a soldered printed circuit board.

If the dimensions are followed closely, alignment may be accomplished by listening to an on-the-air signal near the middle of the two meter band and peaking the capacitor for maximum received signal. If desired, the filter may be followed by an output meter and the capacitor adjusted for maximum power indication. Insertion loss of the filter is less than one dB. (Designed by A.R. Badcock, G8IPQ, England).

Your Checklist for RFI and TVI Problems

What should you do if you are accused of interference? First of all, you should determine if you are the culprit. Check your log or station records to see if you were on the air at the time of the interference. Next, check to see if you are interfering with your own tv set or stereo. If the fault seems to be in your transmitter, follow the guidelines in this chapter.

Ask the cooperation of the complainant. Have someone operate your transmitter and observe the interference on the complainant's equipment. Be courteous and helpful.

If your transmitter is clear of interference-generating emissions, the fault may lie in the complainant's equipment. The next chapter in this handbook covers some of the problems and cures associated with radio and stereo receivers and television sets.

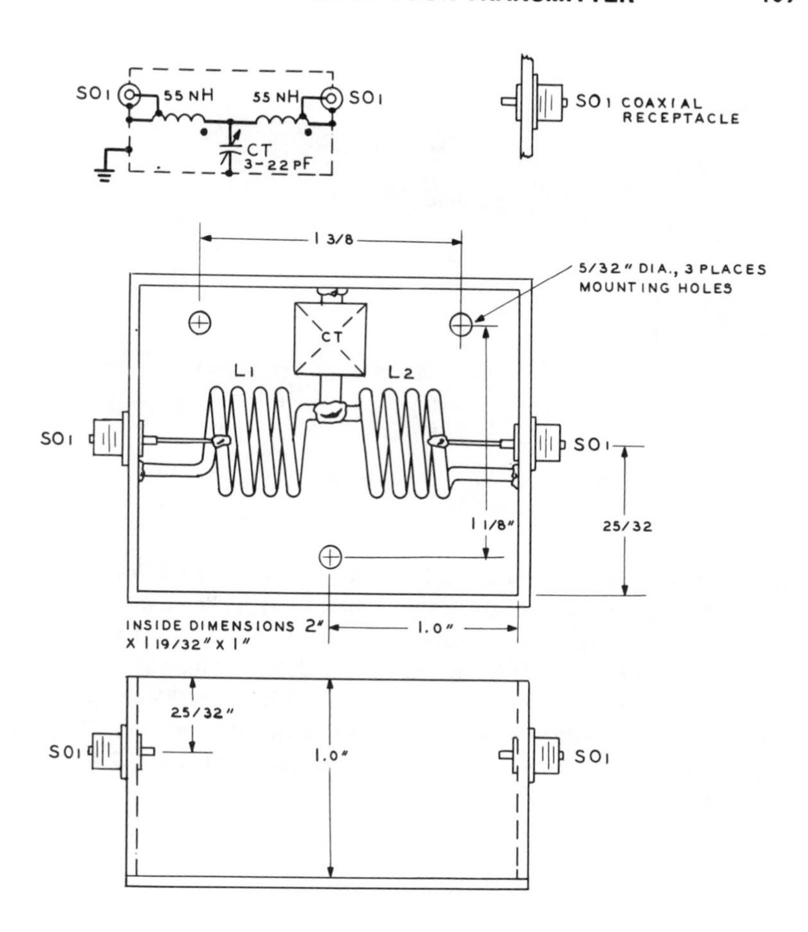

Fig. 5 The lowpass filter for the two meter band. The junction of the coils is suspended freely, supported only by the arm of the variable ceramic trimmer capacitor. The coils are each 4 turns No. 16 enamel wire, 1/2-inch in diameter and 1/2-inch long. Separation between the coils is 1/2-inch. The input and output connections are tapped on the coils about 1/2-turn up from the grounded end.

Filter is aligned by listening to an on-the-air signal near the middle of the two meter band and peaking the capacitor for maximum received signal. Insertion loss of the filter is less than one dB.

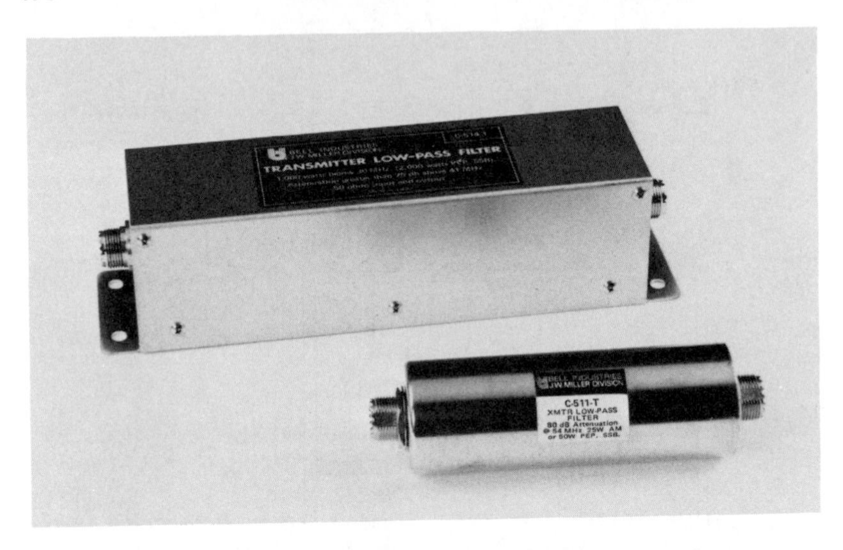

Fig. 6 Representative low pass filters for amateur and CB transmitters. The top filter is useful up to 2 kW PEP power level (1 kW steady state) and provides protection for transmitters operating up to 30 MHz. Attenuation is 75 dB, or greater, above 41 MHz. The filter is designed for a 50 ohm transmission line. The lower filter has a power limit of 50 watts PEP and provides protection for transmitters operating up to 30 MHz. Attenuation is 80 dB, or greater, above 54 MHz. (Photos courtesy of Bell Industries, J.W. Miller Division).

Chapter 10

Eliminating Interference in Home Entertainment Equipment

You can reduce or eliminate many RFI problems in stereo, tv and radio yourself. More complex modifications must be done by a qualified serviceman. This chapter discusses how to solve these problems.

The previous chapter provided do-it-yourself information on solving RFI and TVI problems dealing with harmonics from radio amateur and CB transmitting equipment. It covered steps to be taken external to the equipment to clean up the interfering transmissions. This chapter shows various techniques that can be applied to home entertainment equipment that will help to protect it against unwanted signals. Much of the work external to the equipment can be done by the owner.

The First Steps in Eliminating RFI

If the radio ham or CB operator is fortunate, he may find that cleaning up his transmitter solves the RFI problem. In the majority of cases, however, corrective action must also be taken at the equipment experiencing the interference; the general idea is to prevent the unwanted signal from entering the equipment. Long speaker leads and wires from tuners to amplifiers act as antennas, picking up unwanted signals and feeding them into the sensitive circuits. Signals can also enter via the power cords (Figure 1). Once inside the equipment the offending signal is difficult to eliminate.

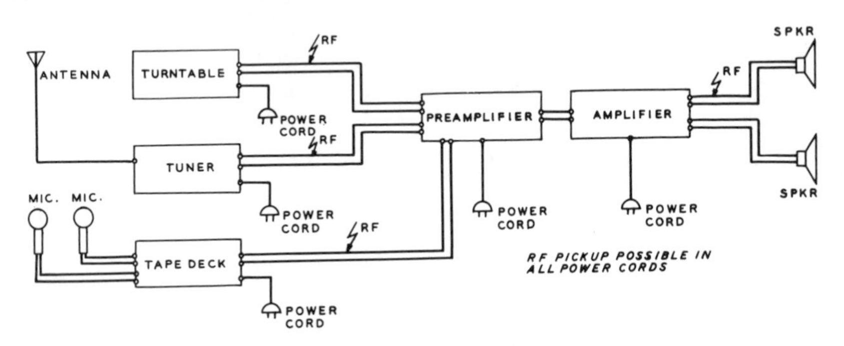

Fig. 1 Input, speaker and power leads can act as antennas, picking up unwanted signals and feeding them into sensitive circuits. Once inside the equipment the offending signal is difficult to eliminate. Shielding and filtering all leads can eliminate offending signals.

Simple corrective measures can be taken, but the addition of RFI suppression circuits in the equipment itself should only be done by a qualified service technician who has the maintenance and repair manuals and expertise in solving RFI-induced problems. More about this later in the chapter.

Remember! Most entertainment devices are covered by a warranty and making unauthorized alterations inside the cabinet voids the warranty. But there's plenty to do outside the cabinet that will reduce rf pickup.

This chapter should serve as a general guide but is not aimed at any specific equipment. It should be referred to as a "road map" as there is no set procedure to follow once it has been determined that the interference is entering via the various cables attached to the equipment.

An Easy Cure for Stereo Interference?

It has been suggested that the first step in investigating interference is to reverse the line plug in the electric outlet. Some stereo gear has the chassis (ground) bypassed to one side of the power line and the polarity of the plug may inadvertently place the chassis above ground as far as interference is concerned. The author has heard of this easy "fix", but has never found a case in which line polarity made any difference in the interference level. However, it is an easy test and it may work for *you.* If it does, it is a temporary cure and should not be accepted as a final solution to the problem.

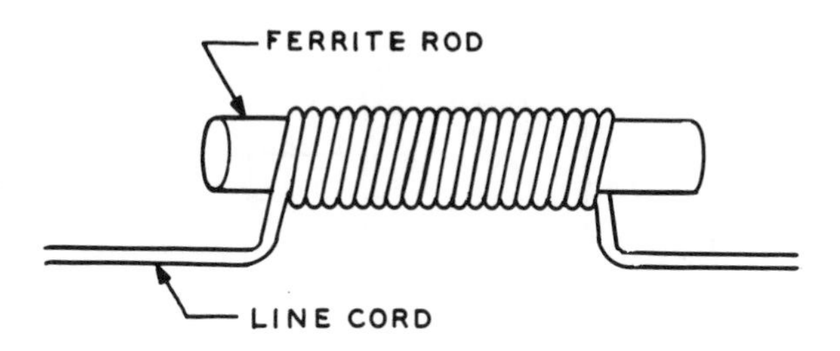

Fig. 2 Wrap the power cord of the stereo or tv set around a ferrite rod, as close to the chassis as possible. This will help choke off any signals arriving via the power cord.

Audio RFI

Audio interference is the detection of rf energy by audio circuits in the equipment. The unwanted signal may be detected by any active device (tube, transistor or IC) in the audio circuitry and the resultant interference voltage is then amplified as a normal audio signal by the rest of the device. Audio RFI is heard regardless of the setting of the amplifier controls or the frequency of desired reception. It seemingly "blankets" the receiver or stereo. The interfering signal may sound quite normal if the transmitter employs amplitude modulation (AM). If single sideband (SSB) is used the voice sounds will be unintelligible. FM and CW transmissions usually produce intermittent volume reduction of the wanted signal but no voice sounds in the receiver.

Before any extensive experiments at RFI reduction are made, wrap one layer of the line cord tightly around a ferrite rod, as close to the chassis as possible (Figure 2). The previous chapters provide information on suitable rods. The coil will help choke off any signals arriving via the power cord. It is a good idea to have a six-foot extension cord pre-wrapped around a ferrite rod to take along with you when you investigate stereo RFI. Should the use of one ferrite coil show an improvement, try using two coils in series to further reduce the interference.

Audio Input and Speaker Leads

Once the power line has been choked off, attention should be turned to the other leads connected to the stereo amplifier. First, disconnect all input leads which include phono leads (two), tape deck leads (two) and tuner inputs (two). If interference has not been reduced it is probable that

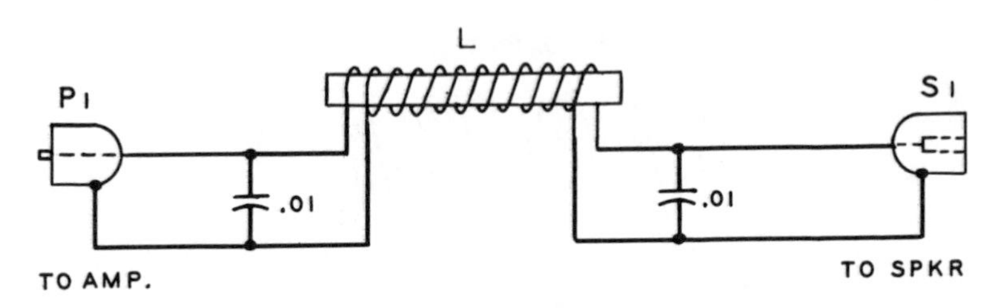

Fig. 3 Lead filter for speaker wires. Capacitors are .01 uF, 600v disc ceramics. Coils are 30 to 40 turns each, wound together on a ferrite rod 5/16-inch in diameter, 4-1/2-inches long. Tape or epoxy windings in place. Wire size should be No. 16, or larger.

the interference is being picked up by the speaker leads.

Lead filtering is an important weapon in the war against RFI. Shown in Figure 3 is a speaker lead filter that can be easily and inexpensively made in the home workshop. One filter assembly is placed in each speaker lead at the amplifier.

The filter is made of a ferrite rod and matching input and output plugs and receptacles that fit the connectors on the stereo equipment. Two lengths of wire are tightly wrapped around the core and taped or epoxied in place. The coils may be covered with heat-shrink tubing if desired. The small capacitors are soldered across the leads on each end of the filter coil.

The leads from the filter coil to the amplifier plug P1 should only be an inch or two long. The leads to the speaker receptacle S1 can be any length. When the speaker lead filters have been placed in the circuit, the stereo should be checked for interference before any input leads are attached. If interference still exists, a more effective line filter must be

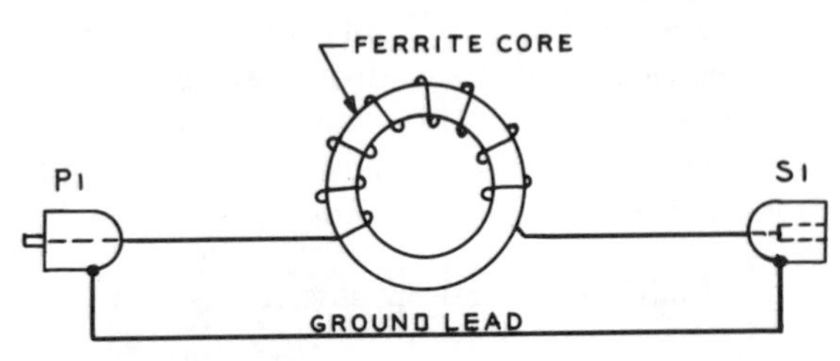

Fig. 4 Miniature filter for phono, tape deck and tuner leads to amplifier. Coil consists of 20 turns No. 20 insulated wire wrapped around a 1/2-inch outside diameter (1/4-inch inside diameter) ferrite core (Amidon FT-50-40, or equivalent). Core permeability is 40. Adapter plugs match stereo fittings.

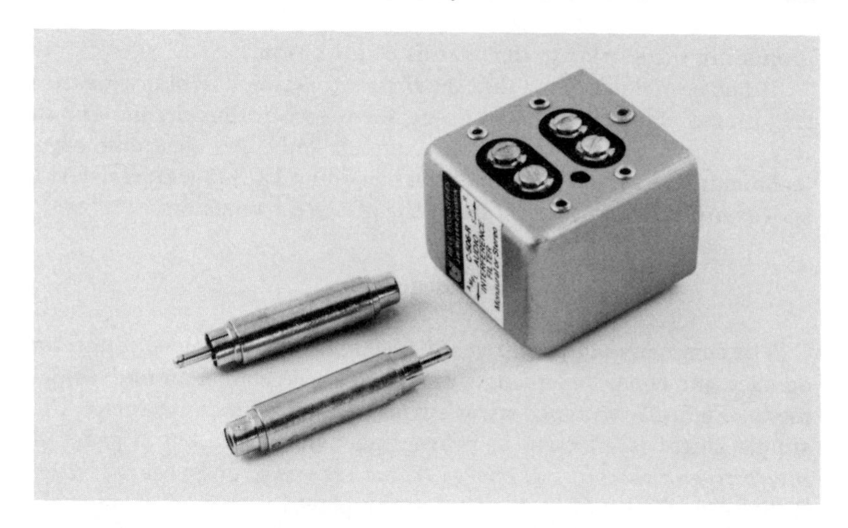

Fig. 5 Stereo input filters are shown at left (J.W. Miller C-505-R). A speaker line filter is at right (J.W. Miller C-506-R).

added to the amplifier such as the unit shown in Chapter 2 of this handbook. It should be placed in the line as close to the amplifier as possible.

The next step is to build up and insert small filters for phono, tape deck and tuner inputs. These small filters are shown in Figure 4. Each filter consists of a small ferrite-core choke and two adapter plugs to match the stereo fittings. The choke coil is placed in series with the center lead between the plugs. Care should be taken to insulate the coil leads so they do not short to the metal cases of the plugs. One filter is required for each input lead. A suitable ready-made filter is the *J.W. Miller C-505R* (Figure 5).

It may be found necessary to wrap the power lines to the tape deck, tuner and turntable around ferrite cores to reduce pickup in these leads. These lead filtering techniques apply equally well to radios, musical instruments, intercom systems, electronic organs and other entertainment devices. In many cases, all the filters are not required.

The Electronic Organ

In most cases, the electronic organ can be protected from RFI by filtering the power cord and placing filters in the speaker leads. In difficult cases it has also been necessary to shield the speaker leads, grounding the shields to the chassis of the organ.

It has been determined that the rf sensitive circuit in many organs is the printed circuit board containing the reverberation circuits and the RFI is heard through the bass channels only. An electronic organ technician can correct this condition by adding 1K, 1/2-watt resistors in series with the base connection of the affected transistors.

Shielded Cables

The commercially prepared shielded cables with molded connectors on each end consist of a single insulated center conductor and a shield made of spirally wrapped wires surrounding the inner conductor. This simple shield is effective in preventing hum pickup but is generally *ineffective in shielding rf energy from the conductor.* In cases of severe RFI the cable should be removed and replaced with a substitute cable made of RG-58/U or RG-59/U coaxial cable of the type used in television or communication service. Suitable fittings will have to be placed on the cable ends to mate with the fittings on the audio equipment. Special, two-wire shielded cable is also available for speaker leads *(Belden 8208,* for example).

In Summary, then........

Unwanted RFI can enter the home entertainment device via the power cable or by the interconnecting leads. Chokes inserted in the leads can block this path. In some instances, the shielded leads must be replaced with leads having improved shielding. Finally, if the interfering signal is exceptionally strong it may be picked up directly within the equipment. In the latter case, shielding and filtering is required within the equipment and that is best left to a qualified service technician.

RFI Data for the Service Technician

RFI-suppression within the home entertainment equipment should be done only be a qualified service technician. Because interference and its solutions are not a part of the curriculum taught in trade schools and junior colleges, most technicians must depend upon the manufacturer's service notes and their own experience to develop cures. The longer the experience, the more numerous and varied are the RFI problems they encounter. Much information on the subject of RFI has been collected by the Consumer Electronics Group of the *Electronic Industries*

Association, a national trade association representing over three hundred U.S. manufacturers of electronic products. The following material is a summary of their service notes concerning RFI and home entertainment equipment.

Interference problems cover home entertainment equipment as well as special installations such as audio visual or public address systems. In all cases, the most common type of interference is due to a phenomenon known as *audio rectification.* This is the detection of modulated signals by the audio circuit. Interfering signals are caused by transmitting equipment but can also be caused by many other types of electrical equipment. The interference usually enters the home entertainment equipment via the power cord or the interconnecting cables. In cases of strong interference, it may be picked up directly by components of the audio equipment. The seriousness of the interference depends upon the strength and location of the interfering source and the shielding properties of the equipment and the building in which the equipment is located. Moving the audio equipment to a different part of the building may or may not help the problem and in most cases it is not feasible. But this simple change may often help solve the interference problem.

Equipment Installation

The service technician should be aware that certain conditions exist which can aggravate an interference problem, such as: 1)- Length of connecting cables between audio components such as phono to receiver, tape deck to recorder, receiver to speakers, etc. 2)- Improper grounding of the equipment. 3)- Insufficient shielding of audio reproducing sources, such as tape playback heads in tape decks, phono pick-ups in record players, and microphones. Cables should be as short as space permits and all equipments should be firmly bonded together by a common ground lead.

The technician should test the equipment in the presence of interference and note if the interference is heard in all modes of operation. He should note if the interference occurs only when touching a certain part, or control, of the equipment and if the interference is heard when the equipment is turned off.

Does the interference stop instantly or gradually when the wall plug is removed? If it stops instantly before the voltage stored in the filter capacitors drops, it is a good indication the interference is coming into the equipment via the power cord.

Does the interference change in intensity or disappear when the

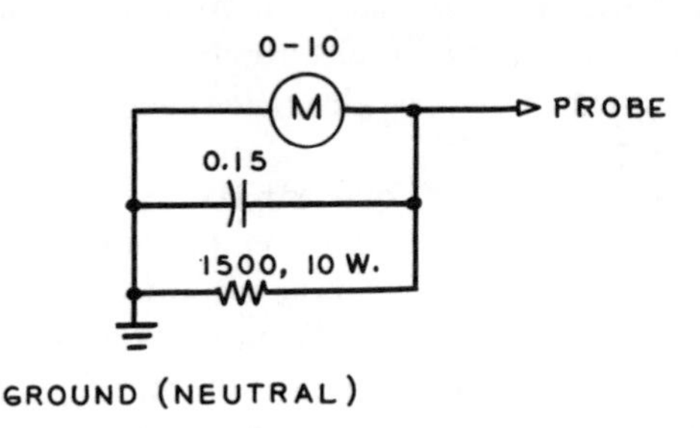

Fig. 6 Leakage tester for home entertainment equipment. The meter is a 10 volt, a-c voltmeter which measures potential between equipment chassis and the electrical ground. Leakage in modern equipment should not run greater than 0.75 volts.

interconnecting cables are removed one by one? This may give a clue as to the portion of the circuit that is interference-sensitive.

Does the interference change in intensity when the volume control is turned up or down? Or when the tone controls are adjusted? Is the interference noted on all channels of the equipment? Armed with the answers to these questions and the Service Manual, the technician is ready to attack the problem.

Safety First--The Leakage Test

Attention to details is the key to success. In the case of a phono arm, if RFI is noticed when the arm is touched, a flexible ground wire between the tone arm and preamplifier ground is necessary. Before the tone arm is grounded a leakage test must be made. If the equipment fails this test, the ground connection may be dangerous to the user and the technician should contact the equipment manufacturer for further instructions.

An a-c voltmeter having 5000 ohms per volt sensitivity, or better, is connected in the test circuit of Figure 6. One side of the meter is connected to a good ground (cold water pipe, etc.) and the probe lead is touched to the metal part of the tone arm, or other accessory of the stereo equipment under test. Measure the a-c voltage as read on the meter and then reverse the a-c plug and repeat the reading. For equipment manufactured prior to 1973, this reading may be as high as 7.5 volts, rms. This corresponds to 5 mA leakage current. A reading in excess of this constitutes a potential shock hazard. For equipment manufactured after January 1, 1973, the

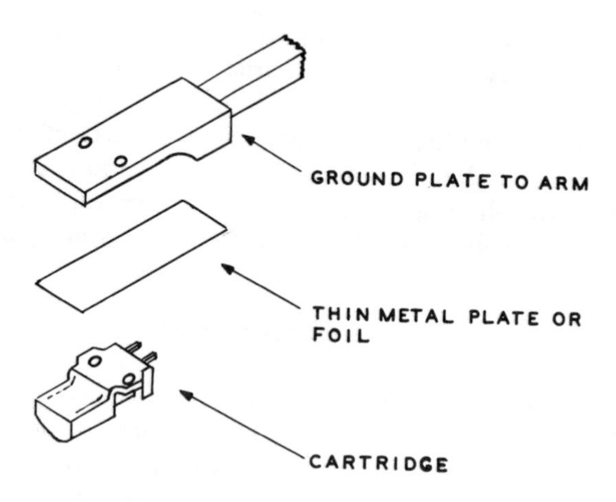

Fig. 7 Metal foil or plate placed between the cartridge and head shell in tone arm will help to prevent RFI.

voltage must not exceed 0.75 volts, rms. This corresponds to 0.5 mA leakage current, the new limit. Any value in excess of this limit indicates a circuit defect which may be potential shock hazard.

Resistance Safety Test

For receivers or amplifiers for which the manufacturer specified resistance tests, comply with the manufacturer's instructions. If these are not available, connect the vhf antenna (if used) and follow this test procedure. Do not plug in the set. Instead, connect both blades of the power plug together and turn power switch to "on".

Using a high-resistance ohmmeter, measure the resistance between the shorted power plug and the antenna terminals, the screws on the back of the cabinet and handle (if any), the shafts of the channel selector and other controls, and any exposed metal parts (escutcheons, overlays, etc.). If any reading is below 600,000 ohms, the cause must be identified and corrected before operating the receiver. In most cases, the reading will show an open circuit except for the antenna terminals which might indicate as low as 5.2 megohms.

The technician, in particular, should test the cartridge and playback arm in any stereo equipment to make sure they are not "hot" to ground.

If the tone arm (as used in this example) passes the leakage test when the equipment is plugged into the a-c outlet, it is safe to add the ground wire mentioned previously.

If the tone arm and head are plastic, a small piece of foil or metal between cartridge and head shell, grounded to the tone arm metal or changer base will help (Figure 7). In the case of a microphone, a ground wire between microphone shell and preamplifier ground is required.

Fire and Shock Hazard

Remember! No modification of any circuit should be made that exposes the technician to a potential hazard. Service work should be performed only after the worker is thoroughly familiar with the safety procedures and instructions in the service information. In performing circuit tests and modifications, the technician should remember these rules:

1- Be sure all components are positioned so as to avoid inadvertent shorts to ground or other components.

2- The job is not complete until all protective devices such as covers, insulators or strain reliefs have been reinstalled per original design.

3- Inspect all modifications and check for poor solder joints, solder splashes, frayed or pinched leads or damaged insulation. Check components for physical evidence of damage.

4- No lead or component should touch a hot tube or other component rated at one watt or more. Lead tension around protruding metal surfaces or edges must be avoided.

5- All critical components must be replaced with exact parts as specified by the manufacturer.

6- When servicing transformerless equipment, always use an isolating transformer to eliminate shock hazard and possible damage to test instruments.

7- Some equipment has a polarized line plug (one wide pin on the plug). Extension cords which do not incorporate this feature should never be used with this equipment.

8- After any modifications and reassembly of the equipment, always perform an a-c leakage test, as described previously. Test all exposed metal parts, antenna terminals and metal knobs to make sure the equipment is safe to operate without danger of electrical shock.

Danger! Implosion and X-ray Radiation

The service technician must be aware that older model tv receivers are equipped with a safety glass in front of the picture tube for implosion protection. Use care when handling the receiver and wear safety glasses. Do not remove the protective glass. All picture tubes used in newer model receivers are equipped with an integral implosion protection system in the picture tube itself.

The technician must also be aware that the high voltage rectifier and regulator *may be a source of X-rays.* These circuits do not emit measurable X-rays when the high voltage is set at the level specified by the manufacturer. It is only when the voltage is excessive that X-radiation is capable of penetrating the walls of the tubes or the high voltage cage and escaping into the room. High voltage should therefore be checked before any work is done on the equipment.

Finally, when working on a tv receiver, all shields must be in place when the set is in operation to avoid electrical shock and possible exposure to X-rays.

Always remember the technician is responsible to the customer for assuring that his equipment is safe after it is modified or worked on!

Interference Correction Suggestions

With regard to stereo equipment, RFI is often caused by long speaker leads. Replacing unshielded speaker cables with shielded ones and/or installing bypass capacitors across speaker terminals will usually cure this problem. Capacitors should be selected with care as too large a value will cause a deterioration in the frequency response of the system, or may cause an oscillation near the upper frequency range of hearing. Most amplifiers can tolerate fairly large capacitors across the speaker terminals before oscillation takes place. It is advisable to connect an oscilloscope across the speaker terminals and check for audio oscillation when placing RFI capacitors in the circuit, or making use of shielded speaker leads. A serious oscillation can quickly rupture the speaker voice coil.

When All Else Fails...

When all of the preceding external corrections fail to reduce the interference to an acceptable level, it will be necessary to go inside the equipment and check for any of the following conditions:

FILTER LOCATION CHARTS FOR RFI PROBLEMS
AM-FM-PHONO-AUDIO EQUIPMENT

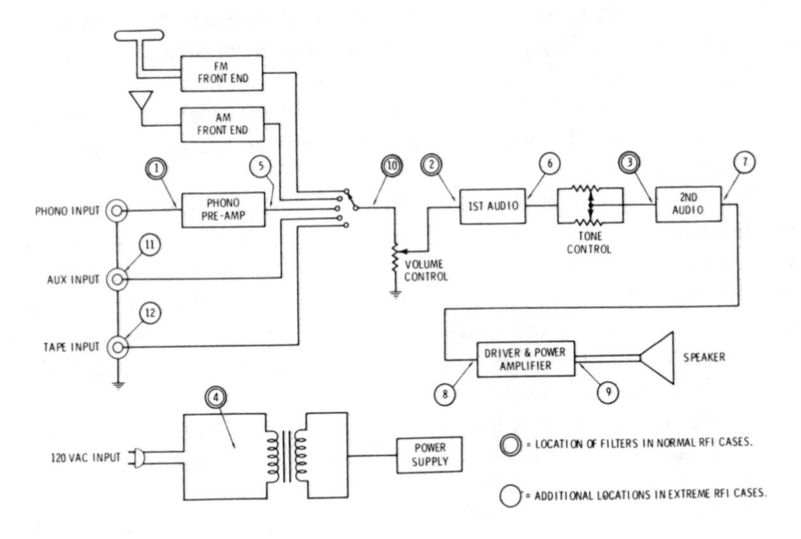

TROUBLESHOOTING CHART FOR RFI PROBLEMS
TAPE DECK, OR THE TAPE SECTION OF AUDIO EQUIPMENT

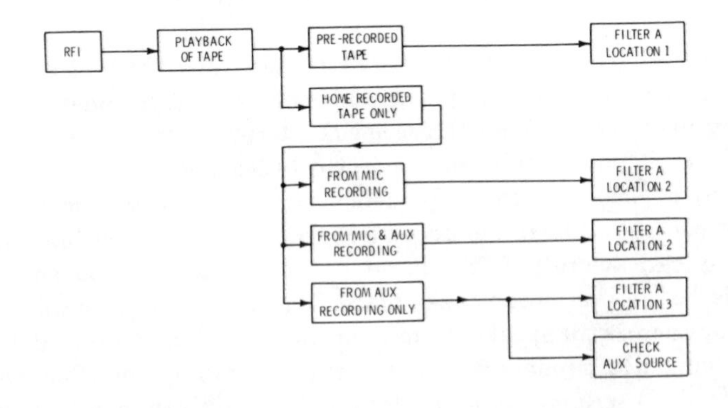

Fig. 8 A filter location chart for RFI problems in am-fm-phono-stereo equipment. Under normal circumstances, only five points (1,2,3,4 and 10) need be considered as critical spots for inclusion of RFI traps. Additional points to be examined in severe cases are noted.

(Chart courtesy of E.I.A.)

—TROUBLESHOOTING CHART FOR RFI PROBLEMS— AM-FM-PHONO AUDIO EQUIPMENT

In the chart below, the type of filters used for certain RFI problems and the points of filter location are given.

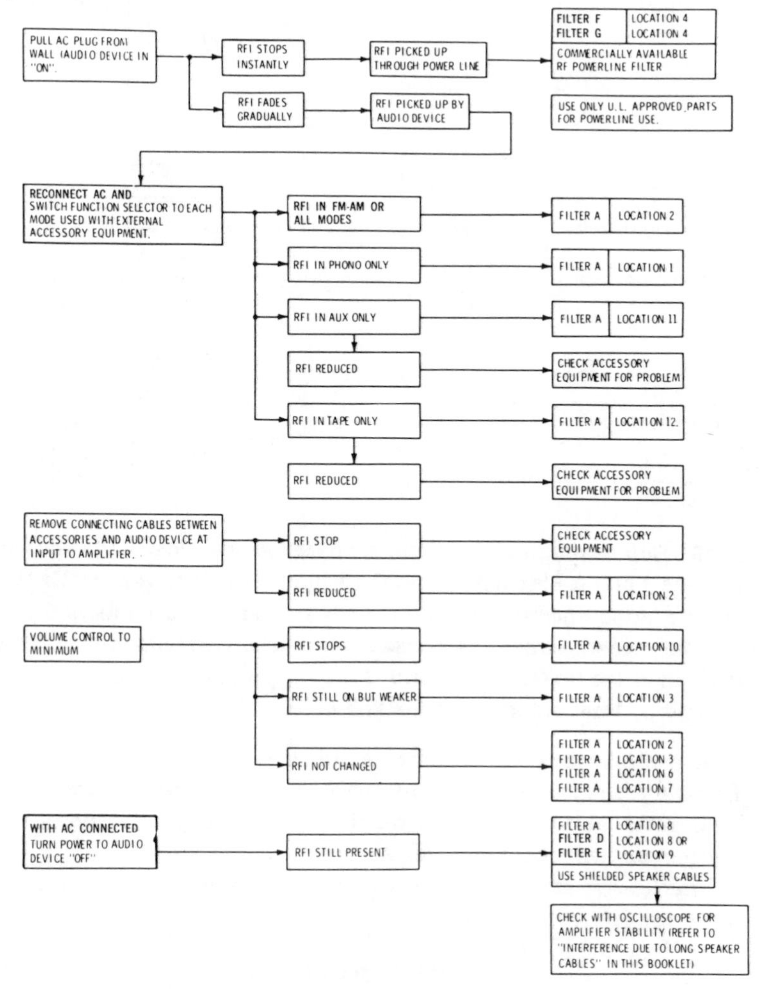

Fig. 9 Flow chart shows sequence of operations that pinpoint the location in the equipment that is RFI-sensitive. Sometimes more than one location requires work to completely eliminate RFI. Representative RFI filters are shown in Fig. 10. These are mounted directly in the equipment at the trouble spot.

(Chart courtesy of E.I.A.)

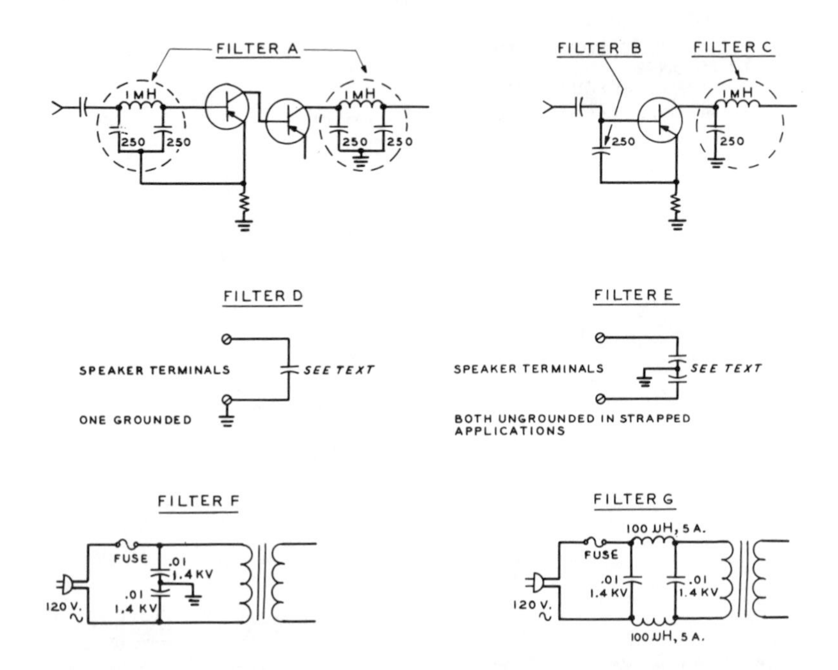

Fig. 10 Representative filter circuits. These filters are installed as outlined in Figures 8 and 9. Filters A, B and C are placed as close as possible to the terminals of the transistor or IC. All leads should be short. Filters D and E are to be placed across the speaker terminals. Capacitance values are discussed in the text. Filters F and G are line filters to prevent unwanted radio energy from entering via the line cord.

1- Bad ground connections and poor solder joints.
2- Electrolytic capacitors that have been in use for several years develop a high internal resistance. Paralleling a new capacitor across the suspected one will check out this source of RFI pickup.
3- If interference still persists, the addition of filters will be necessary. Refer to the following Section.

Internal Filter Circuits

The trouble shooting charts of Figures 8 and 9 indicate where filter circuits can be used to eliminate or suppress RFI. The filter circuits are shown in Figure 10. The trouble-shooter works his way through the charts, following the "road map" which indicates possible locations for filters and the type of filter to be used.

Figure 8 shows the critical points in a typical stereo system where RFI traps may be of benefit. Note that under normal circumstances, only five points in the system (1,2,3,4 and 10) need be considered as critical spots for inclusion of RFI traps. Additional points to be examined in severe cases are noted.

The flow chart of Figure 9 shows the sequence of operations that pinpoint the location in the equipment that is RFI-sensitive. Sometimes more than one location requires work to completely eliminate RFI.

Representative filter circuits are shown in Figure 10. These filters are so small and compact that the filter components may be mounted directly in the amplifier at the trouble spot. In mild cases of RFI the whole filter is not required as a single bypass capacitor placed between the base and emitter circuit may be sufficient.

The Filter Circuits

Various RFI filter circuits are shown in Figure 10. Filters A, B and C are for use directly at the input or output terminal of a transistor, IC or tube. The capacitor is a small ceramic disc unit and the inductor is an encapsulated rf choke *(J. W. Miller 4652-E,* for example). In mild cases of RFI the simple circuits of filters B and C may be satisfactory.

Replacing unshielded speaker cables with shielded ones and/or installing RFI capacitors across speaker terminals will usually cure rf pickup from this source. Capacitors should be selected with care, too large a value may deteriorate frequency response or cause amplifier instability. It is advisable to check amplifier performance for oscillation by means of an oscilloscope connected across the speaker terminals. A strong oscillation can damage the speaker voice coil in a few seconds. The check must be made with the speaker cables connected.

To prevent damage to the amplifier do not operate it with the speaker disconnected unless you are sure the amplifier has a load resistor across the output terminals.

The amount of RFI capacitance the amplifier can stand across the speaker terminals can be determined by experiment, using the oscilloscope as a guide. With no signal input and the amplifier in normal operating mode, there should be no signal pattern on the oscilloscope. The amount of capacitance required to start oscillation is the sum of the capacitance of the shielded speaker cables plus the capacitance of the RFI filter to be added. With the cables connected, the test value of extra capacitance is increased a bit at a time until visual evidence of oscillation is seen. *Once the capacitance at which oscillation starts is known, the total maximum capacitance of speaker leads plus the RFI filter must not exceed one-half this value.* The point of oscillation is constant for an individual amplifier and the length of speaker cable is fixed depending upon the installation requirements of the customer. The capacitance of the RFI filter is the only variable.

Filter Installation Notes

While long connecting cables may act as antennas, it is the audio stages following these so-called antennas which detect the rf signal. These general points must be observed when trying to solve RFI problems:

1- Install the interference filter or filter network as close as possible to the input of the audio device that follows the so-called antenna.

2- Make sure the filter does not significantly change the gain or response of the audio device.

3- Integrated circuits (ICs) are known to be RFI-sensitive. Because of feedback circuits incorporated in IC application, RFI filters should be installed at both input and output terminals of the ICs.

4- Feedback loops in the amplifier can conduct unwanted RFI from the output circuit back into the input preamplifier. Signal pickup by the speaker leads, therefore, can reach the input stages of the amplifier via the feedback path.

5- If RFI originates at the tape deck source only, it must be determined whether it shows up only in playback, or in recording. In the latter case, the RFI can be recorded on the tape along with the program and will show up whenever the tape is played back.

6- Note that in some instances a ferrite bead slipped over a "hot" lead will serve the purpose of an RFI filter. The *Amidon FB-75-101* bead is suggested (permeability of 5,000).

Final Tips

Keep a log of name, model and chassis number when you solve an RFI problem and file the solution to the problem in your copy of the Service Manual. Indicate on the schematic what was done.

Keep a log of the addresses of your customers who experience RFI. It will help you with the next RFI complaint from the same vicinity.

Do not promise the customer a complete cure to his problem. In some cases only a reduction of the interference is possible, and reducing the problem is better than no solution at all.

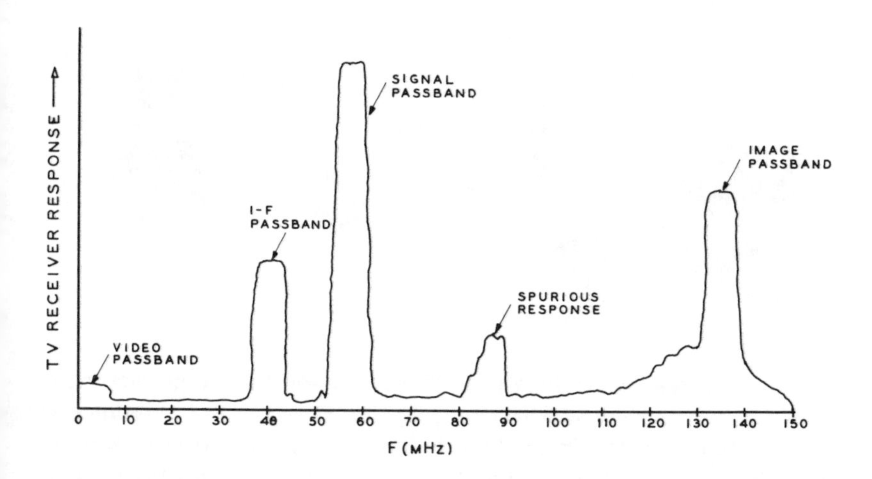

Fig. 11 Representative tv receiver response from 0 to 150 MHz when tuned to channel 2. Note that the image response falls near the amateur 2 meter band. The receiver is also susceptible to interfering signals in the i-f passband at 42 MHz. Responses shown are only relative. The spurious response shown depends upon parasitic resonant frequencies of circuit wiring or components in the receiver

TVI (Television Interference)

FCC studies indicate that the great majority of interference complaints are concerned with impaired television reception. Due to the complexity of the television receiver and the number of audio and video channels within it, very small amounts of rf energy in or near the television channel are capable of causing interference to either picture or sound, or both (Figure 11). Signal response of a representative tv receiver tuned to channel 2 is shown in the illustration. Additional spurious responses may be noted above 150 MHz and up to 600 MHz. For a receiver tuned to channel 11, spurious signal responses as high as 750 MHz may be observed. Thus, the average tv receiver is "wide open" to out-of-band rf energy.

Yes, even the most modern receiver is easily disrupted by an unwanted strong signal, even though the signal is many megahertz removed from the tv channel in use. Such interference is called *fundamental overload* and implies that the signal of a nearby

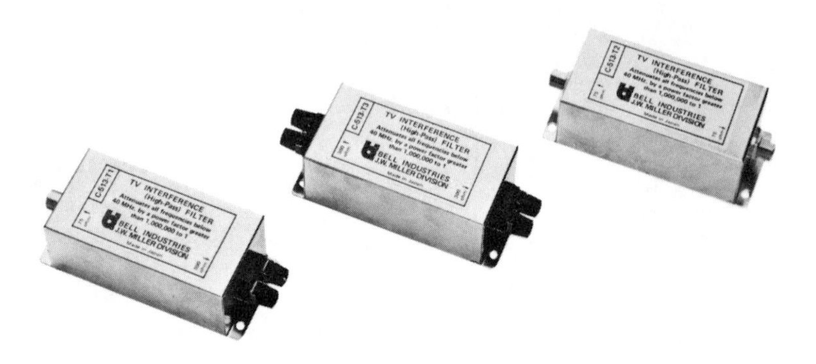

Fig. 12 Representative high-pass filters that will help to protect the tv receiver against overload from strong signals. Filter should be mounted as close to the input terminals of the tv tuner as possible. These filters are made by the J.W. Miller division of Bell Industries. Filters made by Drake, Barker and Williamson and other manufacturers are also available.

transmitter is overloading the front-end circuits of the receiver. This is a very common cause of TVI when the receiver is close to the radio transmitter, particularly with transmitter powers in excess of 100 watts.

Such interference has little to do with the frequency of the transmitter. The symptoms are usually a complete blackout of the tv screen or a dull, white background with intermittent flashes of color that correspond in time with the transmitter modulation. Overload is most commonly noticed on the lower tv channels, particularly 2 through 4.

Receiver Overload and Its Cure

The remedy for overload of a tv receiver (or an fm receiver) is to prevent the interfering signal from reaching the input circuits of the set. This can best be done by means of a *high-pass* filter inserted in the tv antenna lead-in, as near the tuner terminals as practicable. Inexpensive high-pass filters are available for use with the 300 ohm ribbon line used in the majority of installations, or the 75 ohm coaxial lines used with CATV service or with antenna-mounted antenna amplifiers (Figure 12). Purchase the filter which matches the lead-in used with the receiver; the impedance information is printed on the filter label. And don't forget to read the installation information packed with the filter!

Some servicemen mount the high-pass filter inside the tv cabinet next to the tuner antenna terminals. This is an excellent procedure, but is

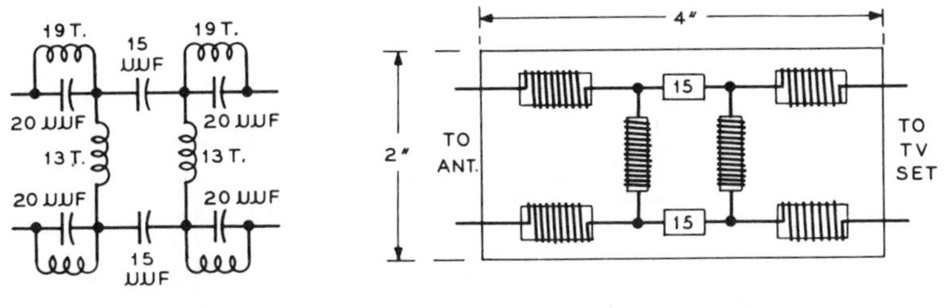

CAPACITORS ARE *CORNELL-DUBILIER TYPE 5W.* 3/6" X 7/16" X 11/16" OR EQUIVALENT SIZE.

WIND 19 TURN COIL ON EACH END CAPACITOR.
WIND 13 TURN COIL ON 10-MEGOHM RESISTOR, 1/4" DIAMETER.

Fig. 13 Homebuilt TVI filter for your tv set. The inexpensive filter protects the receiver from front-end overload. The filter is mounted close to tv antenna terminals.

complicated and time consuming. You should never attempt it on a tv receiver that is not yours. The next best place to put the filter is directly at the antenna terminals on the rear of the receiver. If the filter is contained in a metal box, the box should be grounded to the chassis of the receiver through a .01 uf, 1.4 kV ceramic disc capacitor.

Build Your Own TV Filter

You can build a simple high-pass filter, such as the one shown in Figure 13. It is placed in series with the 300 ohm ribbon lead from the antenna to your television set.

The filter is small and the leads of the components are the only wiring required. The filter is composed of six capacitors and six small coils, interconnected with very short leads. The parts are assembled exactly as shown in the right-hand illustration. A phenolic or plastic board about 4" x 2" is cut, and the complete assembly is fastened to the board by small drops of epoxy cement. The four end circuits are made by winding coils over the bodies of the capacitors. Nineteen turns of no. 24 enamel wire are wound on each capacitor, the turns spaced about the wire diameter. The ends of the windings are scraped clean and the leads soldered to the capacitor wires.

Each center coil is made up of 13 turns of the same size wire wound around a 10 megohm, 1-watt resistor used as a coil form. A 1/4-inch diameter length of wood dowel may be substituted for the resistor, if

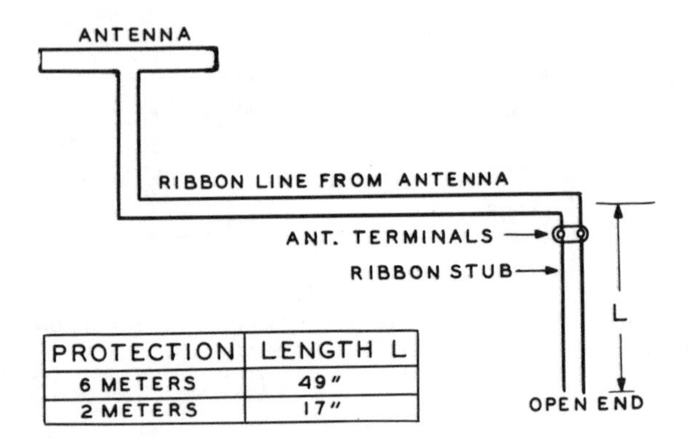

Fig. 14 Inexpensive trap made of tv ribbon line provides protection from amateur signals on the 6 and 2 meter bands.

desired. The turns are spaced the diameter of the wire.

It possible, the filter should be mounted inside the receiver cabinet, positioned as close to the tuner antenna terminals as you can get it.

Receiver Protection from 6 and 2 Meter Signals

Severe TVI can be caused by transmitters operating on the 6 and 2 meter amateur bands. Receiver overload is one form of interference and image interference is another. Either, or both, may occur from a nearby vhf transmitter.

Easily constructed stub traps made of 300 ohm ribbon line will cure vhf TVI in many cases (Figure 14). The strip of ribbon line is attached across the receiver antenna terminals and is cut off 1/4-inch at a time while watching the picture until the TVI is at a minimum. The final lengths will be close to 17 inches for 2 meter protection and 49 inches for 6 meter protection. The free end of the ribbon line is left open.

Tunable traps for 6 and 2 meters are shown in Figure 15. These devices are double parallel quarter-wave stubs taped to the tv receiver feedline near the receiver and tuned with a small capacitor to the interference frequency. The stub is about 40 inches long for 6 meter protection and 11 inches long for 2 meters. Note that the two lines are connected in parallel, one on each side of the tv feedline. Adjust the capacitor with a nonmetallic screwdriver for minimum picture interference.

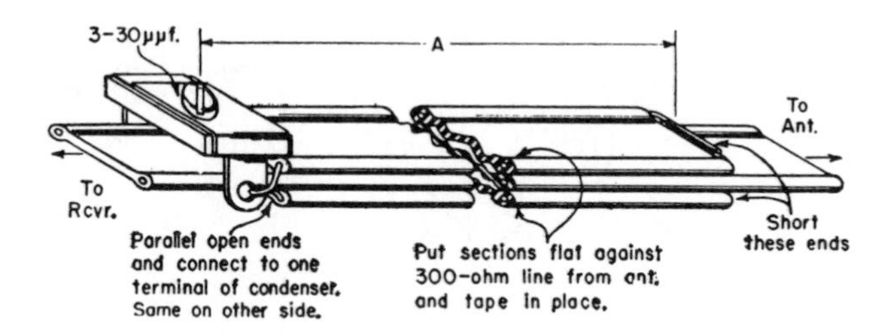

Fig. 15 A tunable trap for 6 or 2 meter interference. Two shorted stubs are connected in parallel and tuned to the frequency of interference by a small compression mica capacitor. Stubs are taped flat against the ribbon feedline.

The stub traps tune quite sharply and must be adjusted carefully. Maximum rejection depends to a degree on the position of the trap along the ribbon line. The trap should be moved along the line until the optimum rejection point is located. The trap is then taped in position. A convenient adjustment technique is to hold the trap temporarily on the line by means of paper tape or several spring-loaded plastic clothes pins. The trap capacitor is adjusted for maximum rejection of the unwanted signal, and the trap then moved along the line for additional attenuation.

Signal Mixing

Nonlinear devices (rectifiers) external to the tv receiver can mix the incoming picture with another strong signal out of the receiver tuning range and produce a new interfering signal within the receiver tuning range. Metal corrosion proceeds uncontrolled in nature and all of the elements required to produce mixing are present in the vicinity of a television set: the maze of pipes, wires and ducts in the home; gutters, fences, guy wires, other radio and tv antennas; sheet metal roofs and power line wiring are but a few of the many readily recognized natural signal rectifiers.

These disagreeable devices can often be located by pounding the walls or floor of a structure or rapping a suspected antenna mast. When these simple techniques fail, a more sophisticated approach is required involving a pickup loop that may be attached to a portable tv receiver (Figure 16). The tuned loop is attached to the antenna terminals of the set via a high-pass filter and used to locate nonlinear devices. The amount of

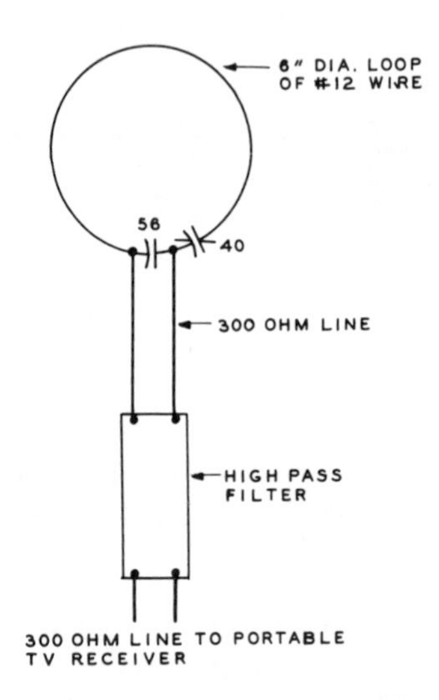

Fig. 16 Tuned pickup loop turns tv receiver into RFI detective. The 40 pF capacitor is a mica compression unit. The 56 pF unit is a silver mica capacitor.

interference on the tv screen provides a clue as to the proximity of the trouble spot. When such a joint is found, it should be separated by insulating material, or shorted out by a copper wire.

Always Check the Antenna System

Regardless of whether a high-pass antenna filter is required or not, it is suggested that the tv antenna system be closely inspected for loose or rusted connections, missing antenna elements and old, faulty lead-in cable. If a preamplifier is in the line, it should be removed from the circuit to see if interference ceases. If it does, a high-pass filter between the amplifier and the antenna is required.

The investigator should finally determine if the RFI is present in the tv receiver with the antenna completely removed. If the RFI remains, it indicates that the circuits of the set are susceptible to RFI and the problem becomes immensely more complicated. The receiver itself must be inspected for proper operation and reference made to the service data-- voltage charts and schematics not normally available to the radio amateur or CB operator.

At this point, the help of a service technician familiar with the receiver is suggested. A television set, or other home entertainment equipment, is a complicated device and the delicate transistors and ICs can be damaged quickly by the uninitiated with inadequate knowledge fooling around with the circuitry. Unless TVI or RFI can be cleaned up by modifications made exterior to the equipment, it is not wise to make alterations or adjustments within the cabinet. This is especially true with your neighbor's equipment. The worst thing you can do is to inadvertently damage his stereo or tv. If a high-pass filter is required, you may supply it, but let the equipment owner install it. You may suggest to him what steps to take, but let him do the work. To touch or otherwise adjust somebody's equipment is folly and can lead to greater difficulties than mere RFI. Be wise! Let the service technician handle tricky problems concerning receiver circuitry.

Sources for Interference Filters

Barker and Williamson, 10 Canal St., Bristol, PA 19007
J.W. Miller Co., 19070 Reyes Ave., Box 5825, Compton, CA 90224
TV High-Pass Filter: Model C-513-T3 (for 300 ohm line)
Model C-513-T2 (for 75 ohm line)
Model C-513-T1 (75 ohms to 300 ohms)
R.L. Drake Co., 540 Richard St., Miamisburg, OH 45342
TV High-Pass Filter: Model TV-300HP (for 300 ohm lines)
Model TV-75HP (for 75 ohm lines)
Radio Shack stores
TV High-Pass Filter: Model 15-1146 (for 300 ohm line)
Model 15-581 (for 75 ohm line)

These filters are available from most equipment dealers. Catalogs may be obtained from the manufacturers.

CATV RFI

Cable tv systems are expanding rapidly in many major urban areas and are posing a potential RFI threat to amateurs. mobile services and viewers. Since they are supposed to be closed (non-radiating) systems, viewers and engineers alike assume they are free of RFI and cause no RFI. Unfortunately, such is not the case.

Many CATV systems utilize the vhf spectrum from 50 to above 225 MHz for their multi-channel signals, providing subscribers with continuous tuning converters to permit them to tune in cable channels outside the standard 12-channel vhf tv range. This places some cable-carried signals into amateur, aircraft and public safety channels. When the system leaks, interference to these services results. Cable leakage is a common occurrence due to corrosion of the cables, loose connectors or cable damage. Many cases of cable system *interference to other services* have been documented.

Other service interference to cable reception is the other face of the problem. Cable subscribers who have paid to watch the special programs being transmitted via cable (but within an amateur band) aren't likely to be very sympathetic when poorly shielded converters or bad connections pick up amateur signals that wipe out their programs.

Potential problems are on the horizon for cable tv. Because cable tv is a regulated utility, cable system operators are required to take care of problems within their systems. To date, this has proven to be an expensive, frustrating and time consuming task.

Computer RFI

In 1980 FCC regulations went into effect that govern the emission of RFI from the so-called "mini-computers" that are becoming so popular in the home. Computers purchased before the FCC regulations went into effect emit considerable interference that may range from a slight fuzziness in the tv picture to a complete blackout of television reception. The amount of interference depends upon the make and model computer, the proximity to the tv receiver and its susceptibility to interference, and the conduction path between the units.

Not only are mini-computers a source of RFI, but all main-frame computers with TTL devices are a source of RFI. The reason is that the switching speed of the microprocessor and TTL devices is in megahertz. The Radio Shack TRS-80, for example, operates at a rate of 1.77 MHz, and emission at that frequency and its harmonics can be heard in a nearby receiver.

Computer noise is mainly radiated by the connecting cables. If the cables and peripheral devices are disconnected, the RFI level will drop. Long leads between the computer and the printer can act as a radiating antenna and should be as short as possible.

Reducing computer-generated RFI requires the same techniques as used with transmitting equipment. The television receiver should have a high-pass filter to prevent harmonics from the computer clock from entering the rf circuits of the receiver. A line filter on the tv set may also be helpful. A power line filter should be placed on the computer. Interconnecting cables should be as short as possible. On some computers, shielded cabling may be used between the processor and the peripheral device to reduce RFI. Unfortunately, many older computers are mounted in plastic cases which permit radiation directly from the unit. Shielded cables will do little good in these instances.

If the computer is in a metal case, the addition of coaxial cables, or twisted-pair shielded cables will be of benefit if the computer case is grounded to a good rf ground.

Some experimenters have made shield sections out of aluminum window screen material and have encased the video monitor in a home-built shield, grounded to the monitor chassis. The shield is large enough to extend up and around the cathode ray tube. All connections to the tube are taped before the shield is fitted in place. Finally, the chassis of the monitor is connected to ground through a .01 uF, 1.6 kV disc ceramic capacitor.

Newer computers are constructed with built-in shields and filters to meet the FCC radiation requirements and do not create an RFI problem.

To check your computer, place a portable tv receiver or a small battery operated broadcast receiver next to it. Moving the radio along the cables will tell you quickly where the interference source is strongest.

As a last resort, moving the computer as far away from the affected tv set may prove helpful. Incorporation of a line filter on the computer, in any event, is a necessity.

The Interference Committee

Many amateur radio clubs have formed an RFI Committee which aids the amateur as well as the listener/viewer plagued with interference problems. This is a volunteer program conducted as a community service.

The committee functions in several ways. In some communities, it has the time and test equipment to check the signal of an amateur operator to make sure his transmissions are free of harmonics or other spurious emissions. It often provides helpful information to the complainant and assists him in obtaining RFI-suppression filters for his equipment. In many cases the committee is the disinterested third party that helps to determine where the responsibility lies by coordinating tests of the transmitting equipment and the complainant's receiver or stereo equipment.

The committee may also provide technical advice to the amateur concerned with an RFI problem and general information about RFI to the community. Some committees have a publicity program which keeps the comunity informed about RFI and its cures by means of newspaper articles and discussions with the governing body of the city. In a few areas, CB operators are setting up RFI committees with the idea in mind of solving and reducing RFI problems arising from CB operation.

Additional information concerning the formation and function of an RFI committee and its relationship with the community may be obtained from the American Radio Relay League, 225 Main St., Newington, CT 06111.

A Final Word About RFI

A burden of RFI falls upon the radio amateur and the CBer as they are a highly visible source of potential RFI to their neighbors. At the same time, they may be qualified to assist their neighbors in understanding and correcting RFI problems. It is a delicate situation fraught with potential problems. But at the same time it may be an opportunity for the amateur or CBer to perform a real public service. It is true that the majority of RFI complaints concerning home entertainment devices would never occur if the manufacturer had included proper shielding and filtering in the device. This is of little concern to the owner bothered with RFI. All he wants is for the interference to stop. The amateur or CB operator can assist in this matter, if only to guide the equipment owner to a good serviceman. Tact, patience and understanding are the key to good neighbor relationships in the face of RFI problems and the burden falls on the amateur or CB operator. He is guilty until proven innocent, like it or not. In an ongoing situation such as this the whole future of amateur and CB communications may be placed in jeopardy. Do your best to be part of the solution, not part of the problem!

Chapter 11

Grounds and Grounding

Ground. What is it? Something to walk on or dig up? Something to fly over? Something necessary and needed in order to properly operate electrical devices? Ground is all of these, and more.

The True Electric Ground

For amateur and CB service the simple rf grounding systems (described in Chapter 9) made up of waterpipe grounds, ground radials and ground rods may suffice. For industrial use and for the more difficult RFI problems a true electric and rf ground is called for. This is expensive and time consuming to achieve but pays big dividends in stubborn cases of interference.

The *true electric ground* is a common reference point in a circuit which is at the same potential as the earth. Earth is literally taken as ground, but not all earth provides a good ground as the electrical conductivity of the earth varies widely depending upon the soil and its moisture content. The best true grounds are the salt water ocean, where conductivity is higher than that of earth, and a salt marsh. The next best ground is the earth itself, especially mineral-bearing soil. The poorest ground is dry, sandy or rocky soil of low mineral content. Many areas of the world have this poor soil and it is thus necessary to simulate a good earth ground.

The efficiency of an earth ground depends upon the resistance, or impedance, of the ground path. If the ground circuit resistance is high, considerable noise voltage may be built up between the earth ground and the point of the equipment that is supposed to be at ground potential.

Ground resistance is made up of the resistance of the ground lead and the ground rod(s) driven into the soil, plus the resistance of the earth-to-rod contact and the resistance of the earth surrounding the rod. The resistance of the lead, the rod, and the rod-to-earth contact are usually insignificant when compared to the resistance of the earth around the rod.

Bureau of Standards tests have shown that if the ground rod is free of paint or grease and the earth is packed tightly around it, the contact resistance is negligible. The resistance of the earth around the rod, however, is not negligible. Earth resistance in the vicinity of the rod can be considerable, but the majority of effective resistance is generally within a radius of six to ten feet of the rod. Beyond that, the area of earth involved in the ground return path is so large that the resistance is unimportant.

Soil composition tests run within an area corresponding to the near-region of a ground rod indicate ground resistance may run from an average figure of about 14 ohms for low resistance, highly conductive soil to as high as 500 ohms for rocky, gravelly soil. It was also determined that the water content of the soil affected ground resistance. For example, a given sample of soil having a moisture content of 10 percent exhibited a resistance of 350,000 ohms per cubic centimeter (350k-ohms/ccm). Increasing the moisture content to twenty percent brought the resistance down to 10k-ohms/ccm. A moisture increase to thirty-five percent reduced the resistance to about 5k-ohms/ccm.

Moisture content of average soil varies from about 10 percent in dry seasons to around 35 percent in wet seasons. This is why the measured resistance of a ground rod will often double from a wet spring to a dry fall.

Effect of Temperature on Soil Resistance

Another item that greatly affects ground resistance is the temperature of the soil. A significant change takes place when ground freezes. For example, the resistance of a soil sample having a stable moisture content rose from 200 ohms/ccm to 500 ohms/ccm as the temperature fell from 70°F to 35°F. When the temperature dropped to 20°F, the soil resistance rose to 6,000 ohms/ccm. At 0°F, the resistance was greater than 40k-ohms/ccm.

This illustrates that frozen soil provides no ground return at all. It is

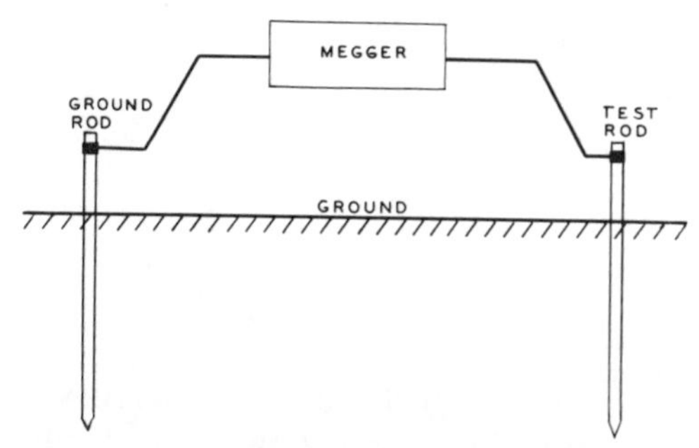

Fig. 1 The "megger" is used to measure ground resistance. This device combines an ohmmeter and voltage source. An auxiliary ground rod is required.

especially important, then, in areas that experience frozen soil that the ground rod be long enough to reach down at least two feet below the frost level. This puts it down into a region that has a permanent moisture level and a stable temperature. Soil at the surface has the most resistivity and is subject to wide variations in resistance with changing weather conditions. Greatest reduction in soil resistance is ordinarily encountered in the first six feet of depth. Most commercial ground rods, therefore, are eight feet long, or longer.

Diameter of the Ground Rod

Weighing these facts, one might think that the diameter of the ground rod might have something to do with a low resistance ground. A comparison has been made between a one-half inch diameter rod and a one-inch rod driven into the earth equal distances. The one-inch rod with four times the surface area decreased the ground resistance only seven percent. In general, then, the rod need only be large and strong enough to withstand the driving force without bending or breaking.

Measuring Ground Resistance

It is possible to measure ground resistance and two techinques are commonly used. The first method is to use a "megger". This is a device which combines an ohmmeter and a voltage source. An auxiliary ground rod is driven into the ground some distance away from the primary ground rod and the megger connected between the rods (Figure 1). The instruction book with the instrument tells you how far away the second

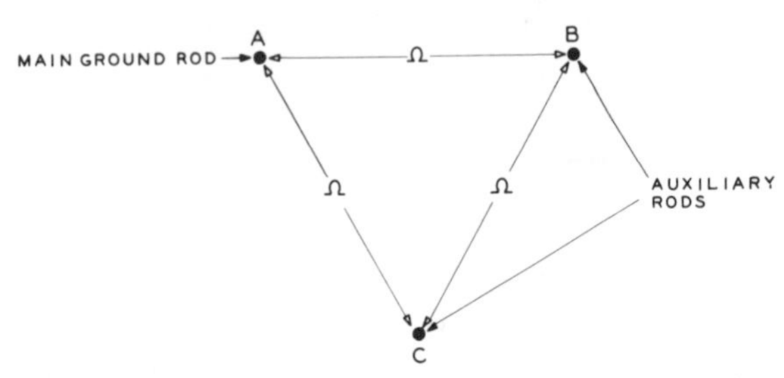

Fig. 2 A volt-ohmmeter can be used to measure ground resistance provided three ground rods are used. Measurements are made between the rods as discussed in the text.

ground rod is placed. The megger, when energized, provides a reference voltage and the ground resistance is read on the dial of the instrument.

A second method of measuring ground resistance makes use of two auxiliary ground rods and a volt-ohmmeter, such as the *Simpson 265,* or equivalent. The two additional test rods are driven in the ground as shown in Figure 2. The rods are labeled *A, B* and *C,* with *A* being the permanent ground rod. Three sets of measurements are taken, between A and B, A and C, and B and C. In each set of measurements, two readings are taken, the second with reversed ohmmeter leads.

For good accuracy, the length of the two auxiliary ground rods should be near that of the one being measured, and they should be located at least 20 feet apart to prevent overlapping of their respective ground resistance areas.

An Actual Resistance Measurement

Refer to Figure 2. The first set of measurements are made between rods A and B. Two readings are taken, the second with reversed ohmmeter leads. The readings are: A to B = 93 ohms
B to A = 67 ohms (reading with reversed leads)

The next set of measurements are between rods A and C.
A to C = 103 ohms
C to A = 71 ohms (reading with reversed leads)

The final set of measurements are between rods B and C.
B to C = 83 ohms
C to B = 113 ohms (reading with reversed leads)

The average reading for each set of measurements is now computed:
Average reading between A and B (67 + 93)/2=80 ohms
Average reading between A and C (103 + 71)/2=87 ohms
Average reading between B and C (83 + 113)/2=98 ohms

The individual ground rod resistances are now computed from the average readings. Each average reading is written as follows:

$$A+B=80$$
$$A+C=87$$
$$B+C=98$$

Solving these three equations, as follows:
Adding equations 1 and 2: $(A+B)+(A+C)=167$

or, $2A+B+C=167$

Subtracting the third equation from the above:

$$\begin{array}{l} 2A+B+C=167 \\ \underline{-\quad B+C=98} \\ 2A \quad = \quad 69, \text{ or } A \quad 34.5 \text{ (ohms)} \end{array}$$

Substituting this value for A in the second equation shows that the resistance of the B rod is 45.5 ohms. Finally, substituting this value in the third equation shows the resistance of the C rod to be 52.5 ohms. The value of the A rod, 34.5 ohms, is the resistance value of the ground rod under test.

Lowering Ground Rod Resistance

A ground rod resistance of 34.5 ohms is considered high and electrical codes in many cities call for a ground resistance in the electrical system of less than 25 ohms. Most power company engineers look for a ground resistance of 10 ohms, or less, in the system.

Steps can be taken to lower ground rod resistance. The ground rod can be extended deeper into the earth. This is done fairly easily if "Copper-Weld" sectional threaded rod sections are used. A section is driven into the earth and a second section is screwed on with a special coupler and the full length is driven into the ground. Power company tests have shown that a ground rod that measured 270 ohms at 8 feet depth measured only 10 ohms at 40 feet.

Another alternative is to drive several ground rods and connect them in parallel. If two extra rods are driven, six feet apart from each other and from the original rod, the ground resistance of the first rod will be cut to about one-third of the original value.

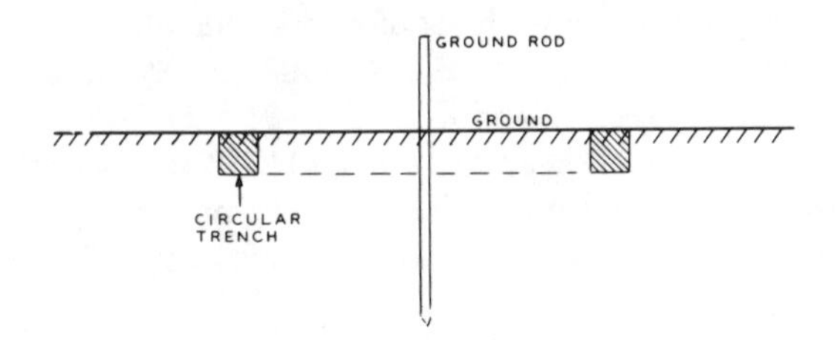

Fig. 3 Chemical ground lowers ground resistance in rocky soil. Trench around the rod is filled with conductive mixture which is kept wet. Common rock salt works well. The trench should have a cross section area of about one foot.

The Chemical Ground

If these methods of lowering ground resistance are impractical, as in rocky soil, the ground around a rod may be chemically treated (Figure 3). A circular trench with a radius of about two feet and a cross section of one square foot is dug around the rod. The trench is filled with an electrically conductive chemical such as magnesium sulfate, copper sulfate, or rock salt which is kept wet. The effectiveness of the chemical ground fades with time and the treatment must be renewed every few years; if not, the ground resistance will creep back to its original high value.

Check Your Ground Connection

Most manufacturers of electronic equipment are remiss in their instruction manuals in that they mention little about grounding except that the equipment "should be grounded". How and what the ground connection should be is rarely mentioned which means the installation may be an electrical hazard and could pose operational problems. Most handbooks, too, fail to discuss adequately the whole problem of grounding and how to achieve a good ground.

As mentioned earlier in this handbook, the most convenient ground connection is a cold water pipe. For shock protection, it is adequate provided a secure connection is made to a metallic pipe. Sometimes the water main from the house to the street is made of plastic pipe, or a plastic coupler is used between the water meter and the service line to prevent electrolysis. This greatly reduces or eliminates the effectiveness of the ground. Another problem is that the water main serving many residences is often no more than a foot or so below the surface of the ground. All of these factors contribute to an inadequate or even dangerous ground.

Some amateurs have used the cold water system of their home or apartment for a radio (rf) ground and have found that the system acts as an antenna. When transmitting, the radio equipment is "hot" and not at ground potential and the imperfect joints in the piping act as multiple signal rectifiers, creating all kinds of severe RFI problems.

In a strong local signal area, the imperfect ground system can pick up broadcast signals and mix them in a manner described earlier in this handbook, creating unnecessary "ghost" signals on the dial of the receiver.

The Ground Lead

An effective ground is at ground potential. A lead from a ground system to the equipment in question, ideally, should have no length at all. Any random length of wire has a measurable impedance between the ends and this value of impedance is placed in series with the ground connection. For example, assume you have a ground lead 16 feet long and your equipment is operating on 20 meters. The lead is a quarter-wavelength long and acts as an impedance transformer; the low impedance ground connection is transformed to a high impedance at the other end of the wire--and the equipment is effectively removed from ground! The wire still serves as an equipment electrical ground as far as the power line circuitry is concerned, but it is a failure as a radio ground.

If the ground lead is 32 feet long at 20 meters, it acts as an effective dipole antenna. It still serves as an electrical power line ground but its effect as an rf ground is exactly opposite to what is required.

The only solution is to make the ground connection lead as short as possible. Then, to improve the rf ground, a quarter-wavelength ground radial wire is added to the equipment along with the electrical ground lead.

An amateur living on the second or higher floors of a building faces a difficult grounding problem. The household electrical ground is adequate for safety but not adequate for rf grounding. Again, a quarter-wave radial ground wire may be the only solution to his problem.

A Practical Electrical and Rf Ground

If you do the job right, it is not difficult to drive an eight foot long ground rod into the ground. A steel post driver can be used, or the ground rod driver shown in Figure 4 can be made up. The driver consists of a lead

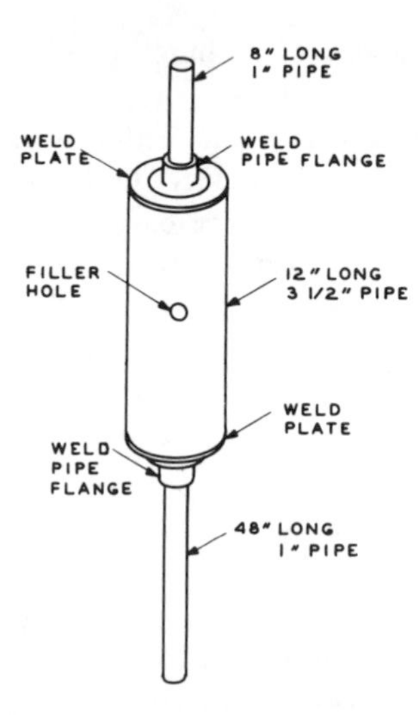

Fig. 4 Concrete-filled drive pipe makes job of driving ground rod easier. Butt end of ground rod is slipped into drive pipe and upper end of pipe is driven downwards with the driver. When the drive pipe cylinder reaches ground level the driver is reversed and short section used as the driver.

or concrete filled cylinder with a driving pipe on one end and a guide pipe on the other.

To start the job, a hole about a foot deep is dug at the position for the ground rod. The bottom end of the ground rod is slipped into the long section of the driver. This prevents the rod from buckling when the going gets tough. The rod is centered in the hole and driving is started with a sledge hammer. Soaking the earth with water makes the job easier. A small ladder or platform is required at first to allow the operator to deliver a sound blow with the driver. Using six inch strokes, the penetration of the rod into the earth is started. Once a foot or two of rod are in the ground, longer strokes can be used. Continue this until the rod is as far down in the ground as it will go.

At this point, the driver is reversed and the short section is slipped over the ground rod. Continue driving until the ground rod is just above ground level. At this point, the connecting wire is attached firmly to the rod which is now driven down below the ground level. This prevents the rod from damaging a lawnmower, and prevents the unwary from tripping over the rod.

Four Copperweld ground rods laid out as shown in Figure 5 make a very effective ground system. The rods are spaced six to ten feet apart

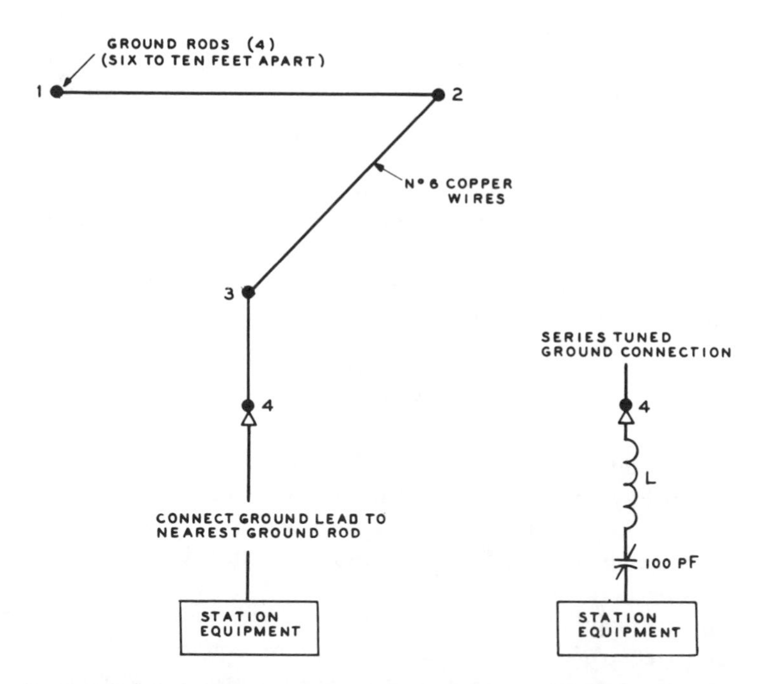

Fig. 5 Parallel-connected ground rods provide a low resistance ground. If the common ground lead to the station equipment cannot be made short, it can be series-tuned to the operating frequency. A capacitor (or capacitor plus inductor) in series with the lead are adjusted to place the equipment at rf ground potential. Coil is 20 turns of no. 12 wire, spaced wire diameter on a 3-inch diameter form. Coil is tapped every other turn for ease of adjustment.

and are connected together with No. 6 house wire.

The distance between the last ground electrode and the equipment to be grounded is critical. The connection should be short and made of heavy wire or copper braid. Connection to the ground rods may be made at any point. The rods need not be laid out as shown, but can be moved about to fit the available space. For best results, the rods should be spaced six to ten feet apart.

If the connecting lead from the ground system to the equipment cannot be made short, it may be series-resonated for operation on any one amateur band (Figure 5). This is accomplished by placing a variable capacitor in series with the ground lead directly at the equipment. The capacitor is adjusted to place the equipment at rf ground. A dc ground may be achieved by placing an rf choke across the capacitor.

The series resonant ground circuit is adjusted to provide the least amount of rf on the equipment. It is a cut and try procedure, adjusting the number of turns in the coil and the value of the capacitance until the symptoms of rf feedback, the "hot" microphone or erratic tuning have been reduced to a minimum. Sometimes the addition of a quarter-wave long radial ground wire at the equipment will improve the situation but occasionally it will enhance the unwanted effects. Experimentation with the tuned ground connection, as shown in figure 5, however, may help solve a "tough" ground problem, especially for those operators living in an apartment, remote from a good ground system.

A Final Word on Grounds

Some amateurs have found to their sorrow that RFI or TVI increases *after* they have made a ground connection to their transmitter. This is an indication that the ground is imperfect and that rf currents may be flowing through the ground circuit from the transmitter to the television receiver or stereo equipment.

The only solution to this perplexing problem is a complete study of the ground system of the whole residence. Insofar as plumbing is concerned, it may be necessary to ground the pipes by means of ground rods at several points. It may be necessary to decouple the electrical wiring in the walls of the house from the transmitting antenna. This can be done by placing a .01 uF, 1.4 kV disc ceramic capacitor across the prongs of an electric plug and plugging it into wall receptacles at random to note if the interference level decreases. Several such bypass plugs may be necessary. A combination of multiple ground rods and bypass plugs will be a great help in difficult cases of RFI when transmitter and receiver are located in the same building.

Chapter 12

Vehicle Noise Suppression

One of the greatest and most useful creations of man is the gasoline engine. It is the universal power source for land and marine vehicles, lawn mowers, snow blowers, chain saws, electric generators and countless other devices in daily use about us. It is also without doubt the greatest source of RFI seen or heard on receivers today (Figure 1).

Vehicles manufactured in the United States are RFI-suppressed at the factory to provide satisfactory reception on the factory installed car receivers; in addition, they must conform to SAE (Society of Automotive Engineers) standards of suppression for external receivers in adjacent vehicles. However, as the vehicle ages, and is not given the proper tune-ups and maintenance, the RFI generated by these engines grows steadily worse and can seriously affect reception, especially in nearby two-way radio mobile units used by CBers, amateurs, police and fire units and marine services.

In addition to the ignition noise generated by the gasoline motor, vehicles also contribute noise generated by their electrical circuits and additional radio noise may be created by the movement of the vehicle through the atmosphere.

Vehicle noise suppression can be attacked in two ways; the first is the discovery and suppression of the noise at the source and the second is the addition of noise limiting or suppression circuits in the receiver in the vehicle. Both methods may be necessary in some circumstances.

The vehicle electrical circuits that generate RFI cannot be modified very much because that would drastically reduce the efficiency of the engine and because of the adverse effect of chemical pollution from poorly burned fuel mixtures. Thus, automotive RFI can be reduced to a

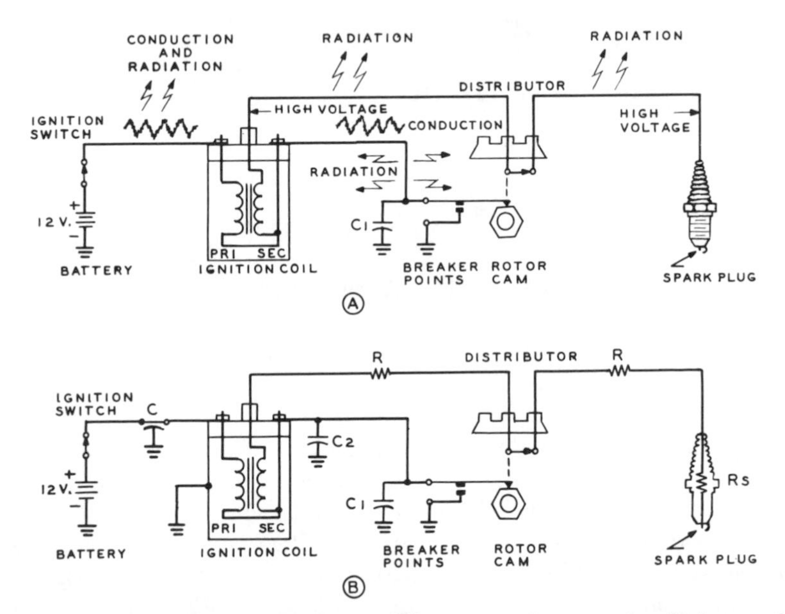

Fig. 1 The gasoline engine is a prolific source of radio noise. Noise can be radiated and conducted from the primary circuit directly into the radio equipment. Radiation from the high voltage secondary circuits can reach the radio equipment by radiation from the wiring and components. Illustration B shows RFI suppression by the addition of coaxial capacitors (C), high voltage ceramic capacitors (C1 and C2) and suppression resistors (R). Special spark plug with internal suppression resistor is also used.

degree that is tolerable but cannot be totally eliminated.

The radio noise can enter the communication equipment via the antenna by direct radiation from the noisy circuits or it can be conducted in by way of the power wiring. Many times the noise comes in both ways.

Checking Out the Vehicle

After installing your communication equipment and before taking any RFI suppression measures, operate your equipment with the car in motion to see how much radio noise your vehicle generates. You may be pleasantly suprised that it produces little, if any, interference. Most amateur and commercial vhf FM transceivers, for example, are relatively immune to vehicle noise and work perfectly without any RFI suppression. On the other hand, CB transceivers which are AM

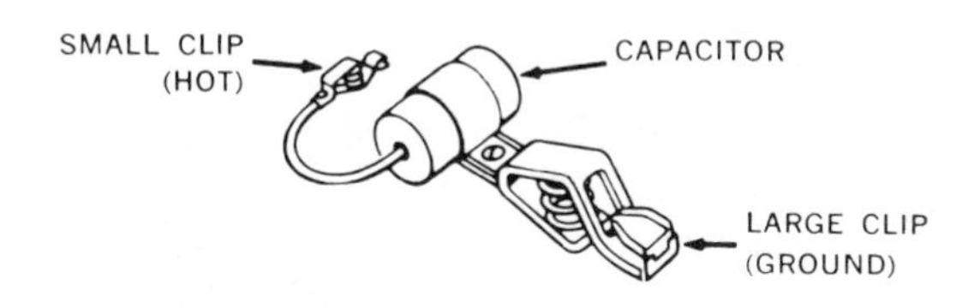

Fig. 2 Clip-on coaxial capacitor is used to test primary circuits for RFI.

(amplitude modulated) and some SSB equipment are far more likely to suffer from automotive noise.

Identifying Interference

Each type of interference you hear on a mobile receiver gives a clue as to its identity by its characteristic sound. *Ignition noise* is identified by a popping sound that increases in tempo with higher engine speed. It stops instantly when the ignition key is turned off at fast idle. *Generator or alternator noise* is a high pitched musical whine that increases in frequency with higher engine speed. It does not instantly stop when the ignition key is turned off at fast idle. *Voltage regulator noise* is a ragged, rasping sound that occurs at an irregular rate. It does not stop instantly when the ignition key is turned off at fast idle. *Instrument noise* is a hissing, crackling, clicking sound that occurs irregularly as the gauges operate. It is usually worse on rough roads. A loud, intermittent, harsh noise may be heard from the voltage limiter used with fuel and temperature gauges. It may change in intensity when the dash is jarred. *Wheel and tire static* is an irregular, popping or rushing sound that occurs only in dry weather at high speeds. It disappears when the brakes are lightly applied.

Tracing Interference

If the radio noise is elusive, or the source not clearly defined, you may be able to locate it with a special tracing technique that will save you time and effort when you finally apply interference suppression techniques. You will need a clip-on *coaxial capacitor* (Figure 2) and a "sniffer coil" (Figure 3). The capacitor is clipped to ground (the engine or vehicle frame) and the small clip lead touched to all live electrical connections in the battery and alternator or generator circuit. (Note: *Do not* connect to

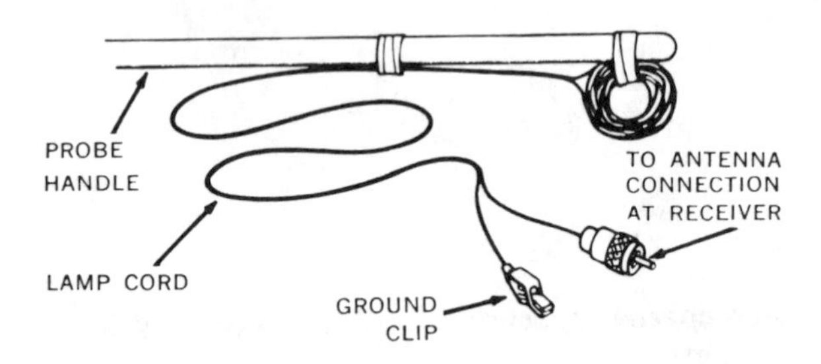

Fig. 3 "Sniffer coil" is ten turns of insulated hookup wire about three inches in diameter taped to end of a wood probe handle. A length of lamp cord connects one end of coil to vehicle ground and the other end to the center conductor of the antenna receptacle on the radio receiver.

the field terminal on an alternator or generator). If the noise level in the radio drops, it indicates the particular circuit must be equipped with suppression devices.

To use the "sniffer coil", attach the plug to the receiver in place of the regular antenna. The clip is grounded to the auto frame or engine. Start the engine and turn on the radio. Probe around the engine and wiring with the coil. Bounce or shake the vehicle during probing. Maximum interference will be heard when the probe is near the noise source.

The next thing to do is to place a dummy load on the receiver antenna terminals in place of the regular antenna feedline. Start the vehicle, turn on the radio, and listen. You won't hear signals, of course, but you may hear noise. If you do, the noise is entering the radio via the power and control cables.

The Coaxial Capacitor

One of the devices that is most effective in suppressing vehicle radio noise is the *coaxial capacitor* (Figure 4). This device will pass noise current to ground without short-circuiting the direct current circuit, which passes through the capacitor from one end to the other. Conventional bypass capacitors are not very effective at the higher frequencies and hf or vhf radio communication equipment make the use of coaxial capacitors necessary to reduce radio noise.

To place the capacitor in a circuit, the body flange of the capacitor is bolted to the engine frame or body of the vehicle. Make sure a good electrical contact is made at this point. The lead to be filtered is broken and the free ends are connected to the two capacitor end terminals. The

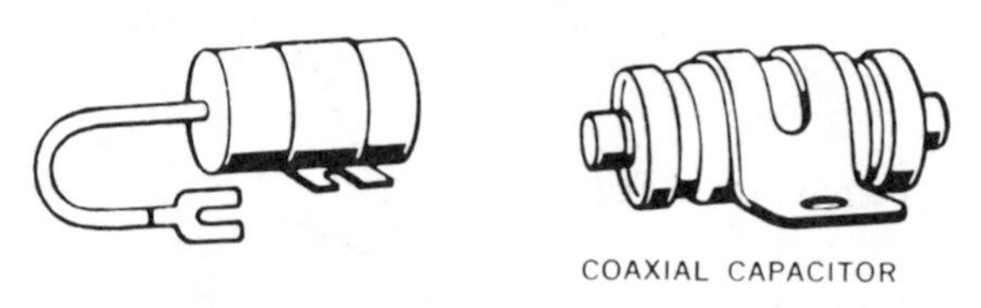

Fig. 4 Conventional bypass capacitor (left) is satisfactory for broadcast reception but useless for shortwave or vhf reception as it does not bypass all radio noise to ground. Coaxial capacitor (right) is effective at all radio frequencies. Frame of capacitor is grounded and lead passes through the body of the capacitor.

coaxial capacitor is rated for the number of amperes it can carry between the two terminals. Make sure the one you use is robust enough for the job.

Ignition Noise Suppression

The ignition system (Figure 1) furnishes a high voltage spark to ignite the gas-air mixture in the cylinders of the engine. About 8,000 volts is required for this purpose. The distributor breaker points select the voltage for the proper plug and an interrupted dc voltage is provided to the ignition coil by a separate set of breaker points driven by the engine. To reduce the radio noise generated by this system, it is necessary to install noise filter capacitors on the ignition coil and to restrict radiation from the wiring. This reduces the transfer of noise to the radio equipment by both conduction and radiation.

First, remove the coil and scrape the paint from the brackets and mount at the mating surfaces. Bolt the coil back in place using lock washers under the nuts to achieve a firm ground connection. Next, install a .005 uF, 1.6 kV ceramic disc capacitor at the coil *distributor* terminal and solder the free lead to the mounting bracket. Finally, install a 0.1 uF coaxial capacitor near the *battery* terminal of the coil (Figure 5). This is connected in the line from the ignition switch. Do not use a conventional bypass capacitor as it is not effective in the hf-vhf radio bands.

Once the coil modification has been made, the level of radio noise should be checked with a dummy antenna connected to the radio equipment. If the noise can still be heard, a coaxial capacitor must be installed on the "hot" power lead to the radio and bolted to the radio case.

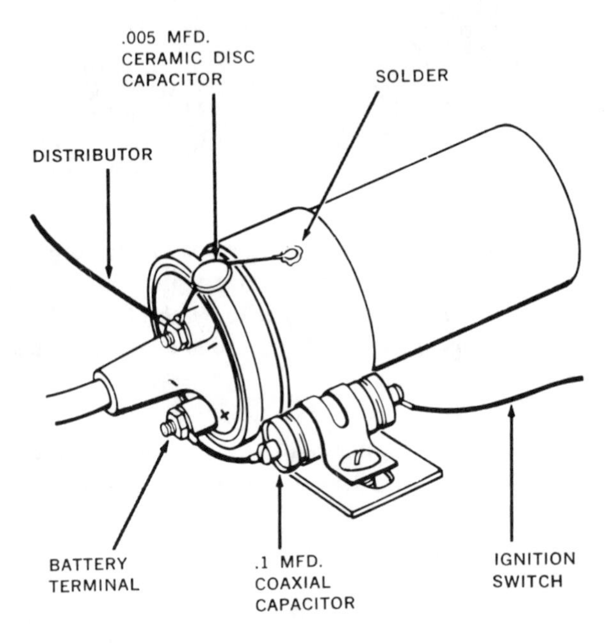

Fig. 5 Noise capacitors placed on the ignition coil help suppress RFI. Ceramic disc capacitor is placed on the distributor terminal and coaxial capacitor is placed in series with wire to ignition switch. Coil case is grounded to vehicle.

Radiation Noise

Attention should now be turned to the high voltage wiring which can radiate ignition noise directly to the antenna of the communication equipment. Noise suppressor resistors are commonly used to reduce the level of noise radiated by the spark plug wiring. Various types of resistors are available. Some are separate components for use at the distributor or spark plug terminals. Sometimes a suppressor is moulded into the distributor rotor. Suppressors are often original equipment on foreign cars but in the U.S. are mainly service items at additional cost.

The most popular form of suppressor resistor is resistance ignition cable which contains a resistive conductor instead of wire. These cables are generally available from automotive service/sales outlets.

A more effective and expensive device for noise suppression is the resistor spark plug (Figure 6). This is a special plug with a built-in

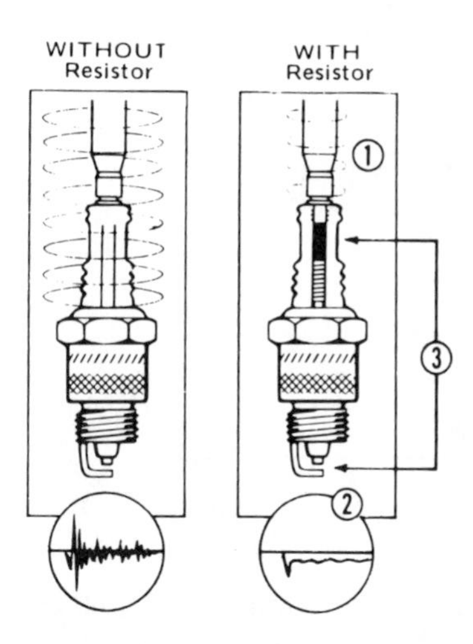

Fig. 6 Resistor spark plug has built-in noise suppressor (3). The effectiveness of the plug is due to the proximity of the resistor to the spark gap (2). Radiated noise from wiring (1) is greatly reduced.

resistor. The effectiveness of the plug is due to the proximity of the suppression resistor to the spark gap which prevents the radio noise from escaping from the plug.

In severe cases of radiation noise both resistor plugs and suppressor cable must be used. The plugs and cable are often combined in a shielded ignition kit (Figure 7). This add-on package includes a shielded enclosure for the distributor cap and the top of the ignition coil, a coaxial filter capacitor for the battery wire, and a set of shielded ignition cables to run from the distributor to the spark plugs. Some kits include shields for the plugs. The kits must be purchased for a specific engine.

Shielded ignition wiring may affect engine performance as the voltage delivered to the spark plugs is considerably reduced because of the capacitance of the cable to ground; the kit designed for your car's engine provides the maximum voltage possible. The shielded ignition system must be maintained in top condition or the performance is significantly reduced. The spark plugs should be re-gapped and serviced at frequent intervals. In some cases, an ignition coil of higher output must be installed to maintain engine performance.

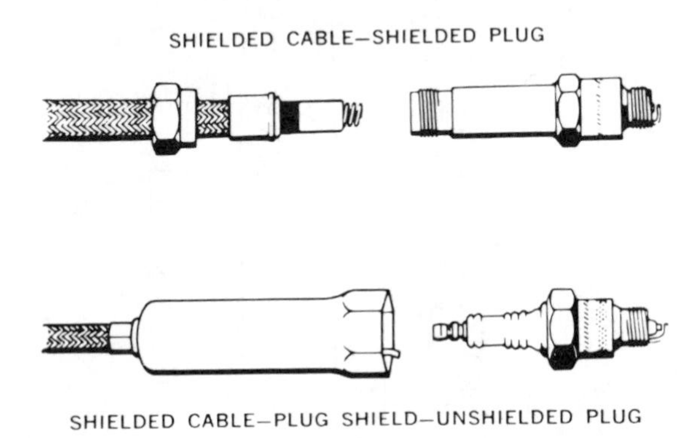

Fig. 7 Resistor spark plug and shielded ignition cable are used in severe cases of RFI noise. This add-on package is available in a kit and must be purchased for a specific engine.

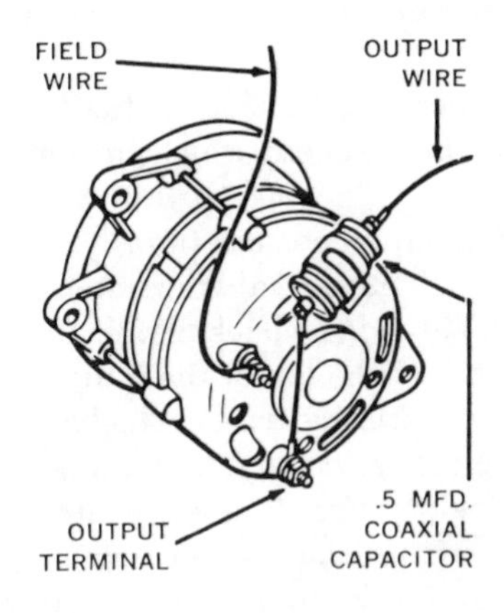

Fig. 8 Alternator "whine" is suppressed by a coaxial capacitor placed at the output terminal and attached to the alternator frame. Two capacitors are required for dual terminals of heavy duty alternator. Make sure the slip rings are clean and that brushes are making good contact.

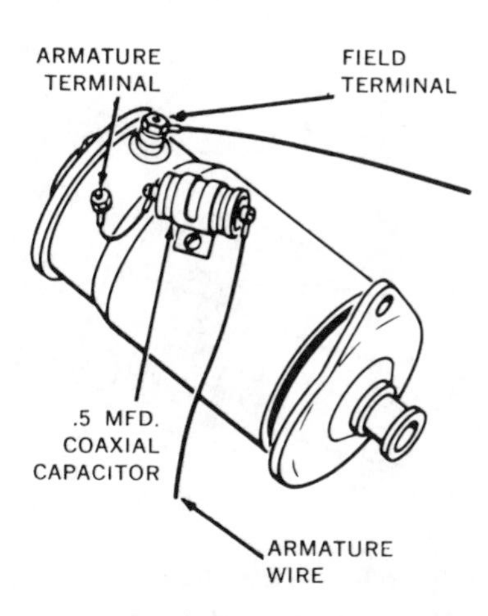

Fig. 9 Generator commutator should be cleaned and checked. Coaxial capacitor is placed in armature lead. Field terminals should not be bypassed. Make sure the body of the generator is properly grounded.

Alternator or Generator Noise

To prevent alternator "whine" from affecting the communication equipment, clean the slip rings and make sure the brushes are making good contact. Then install a 0.5 uF coaxial capacitor at the output terminal of the alternator. Ground the capacitor to the alternator frame (Figure 8). Two capacitors are required for the dual terminals of a heavy-duty alternator.

If the vehicle is equipped with a generator, the commutator should be cleaned and checked to make sure the brushes are seated properly. The factory installed filter capacitor (ineffective in the hf-vhf range) is removed from the armature (A) terminal and replaced with a 0.5 uF coaxial capacitor. Do not connect a capacitor to the field terminal (F) of the generator, (Figure 9). Finally, to make sure the body of the generator is properly grounded, install lock washers on the mounting bolts between bolt head and generator and between nut and vehicle body.

Voltage Regulator Noise

Little or no RFI is caused by the voltage regulator on new vehicles equipped with a solid state ignition system. The older mechanical

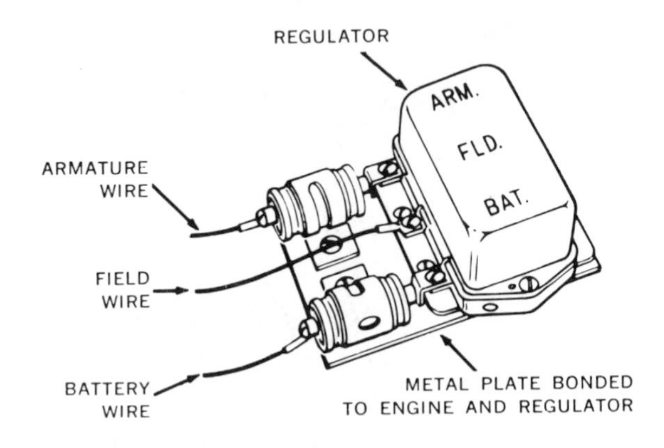

Fig. 10 Regulator is bolted to a metal plate bonded to engine. Armature and battery leads pass through coaxial capacitors bolted to the plate. The field wire is unbypassed. In severe cases of interference, the field wire can be bypassed to ground with a .002 uF capacitor placed in series with a 4 ohm, 1 watt resistor.

regulator, still in general use, can produce a crackling noise when it is operating. The noise is caused by sparking at the regulator contacts. It is an intermittent noise as it only appears when the regulator is working. To reduce it, a 0.1 uF coaxial capacitor is placed in series with the battery lead (B) directly at the regulator, with the capacitor case grounded to the frame of the regulator. A second capacitor is placed in series with the armature lead (A). The field lead (F) cannot be bypassed in this manner without damage to the regulator. Instead, a .002 uF ceramic disc capacitor is placed in series with a 4 ohm, 1 watt resistor from the field terminal to regulator frame. Figure 10 shows a regulator installation.

Instrument and Accessory Noise

The gas gauge, oil gauge and temperature gauge and various "idiot lights" sometimes require RFI suppression. The most practical way to determine if these circuits are noisy is to disconnect all of the "hot" leads to the gauges and reconnect them one at a time while listening for a change of noise level in the receiver. This rasping noise can be suppressed by installing a "hash choke" and a 0.1 uF coaxial capacitor at the point where the gauge is mounted and grounding the capacitor case to the frame of the car (Figure 11). A similar capacitor may be required on the lead to the windshield wiper motor, heater motor or electric window or

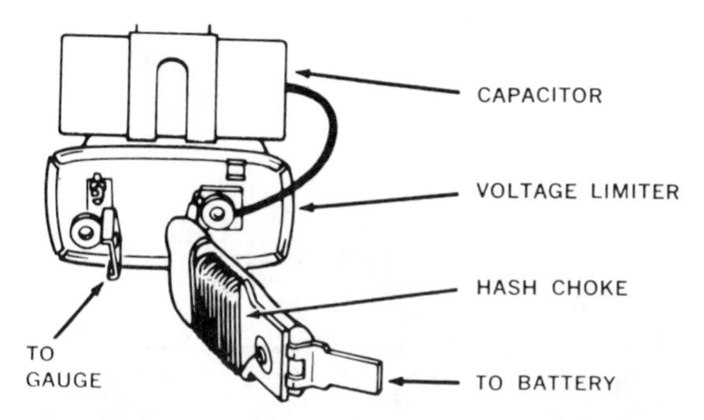

Fig. 11 Instruments and "idiot lights" sometimes require noise filtering. Rasping noise of gauges can be suppressed by installing a "hash choke" in series with the lead and a coaxial capacitor at the point where the gauge is mounted. Ground capacitor case to the frame of the vehicle.

seat motors. In some cases, where motor noise is light, a .01 uF ceramic disc capacitor at the motor terminal to ground will do the job.

Bonding

It may be necessary to bond together body parts to establish a ground system common to every part of the vehicle. Any piece of metal that is ungrounded can pick up ignition noise in the same manner as an antenna does and radiate it to the antenna of the receiver. The car exhaust system is a prime offender. It is connected to the engine which is alive with radio noise and acts as a good antenna because of its length. The engine is mounted on rubber engine mounts to reduce vibration and the exhaust pipe is hung by corded rubber hangers. The whole system, then, acts as a noise antenna. A metal bonding strap is installed across the engine mounts between the motor block and the frame and a bonding strap is installed from the exhaust pipe in front of the muffler to the frame. A similar strap should be installed between the exhaust pipe and the frame. Additional bonding straps from the firewall to the engine block will help in stubborn cases. A bonding stap from the frame of the alternator to the frame is also a good idea.

Other bonding points are shown in Figure 12. The hood and trunk lid should also be bonded. Place a flexible, braided strap across each hinge and install metal clips that ground the large metal surfaces at several points opposite the hinges. This can be done by scraping the metal bare in a few spots and making a pad of braid material that will be pressed in

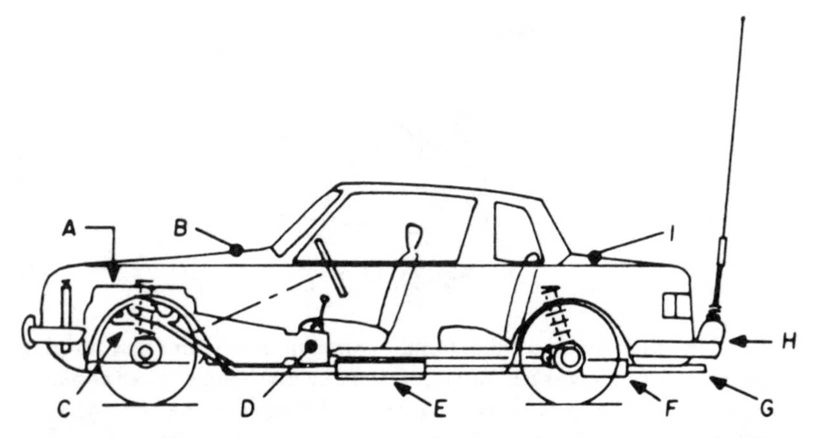

Fig. 12 Points on the vehicle that might require RFI bonding. A-Bond engine to frame. B-Bond hood at hinges. C-Bond across shock abosrbers and rubber engine mounts. D-Bond drive train assembly to frame. E, F, G-Bond exhaust pipe, muffler and tail pipe to frame. H: I-Bond bumper and trunk lid to frame.

place by the sponge rubber weatherseal strip. If your vehicle has body-and-frame construction, place bonding straps across the shock mounts that are used between the body and frame members.

When installing bonding straps, scrape away the paint, sand the area of contact and then apply conducting grease (Penetrox) to both the bond and the body. Attach the bond with bolts or a heavy self-tapping screw and lock washer. U-clamps should be used on the exhaust pipe.

Making a Bonding Cable

A good bonding cable can be made from the outer shield of old RG-8/U coaxial cable. The shield is removed and flattened. After it is cut to length, a heavy soldering iron is used to solder the braid ends into a solid, flat mass that can be easily drilled for a bolt or screw. The flexibility of the braided bonding cable prevents it from breaking under vibration and flexing. Sometimes it is possible to find bonding straps at an automotive supply store that will do the job.

All of this sounds like a lot of work---and it is. Happily, most of it will not be necessary and you will not have to do a complete job of filtering, shielding and bonding on your vehicle.

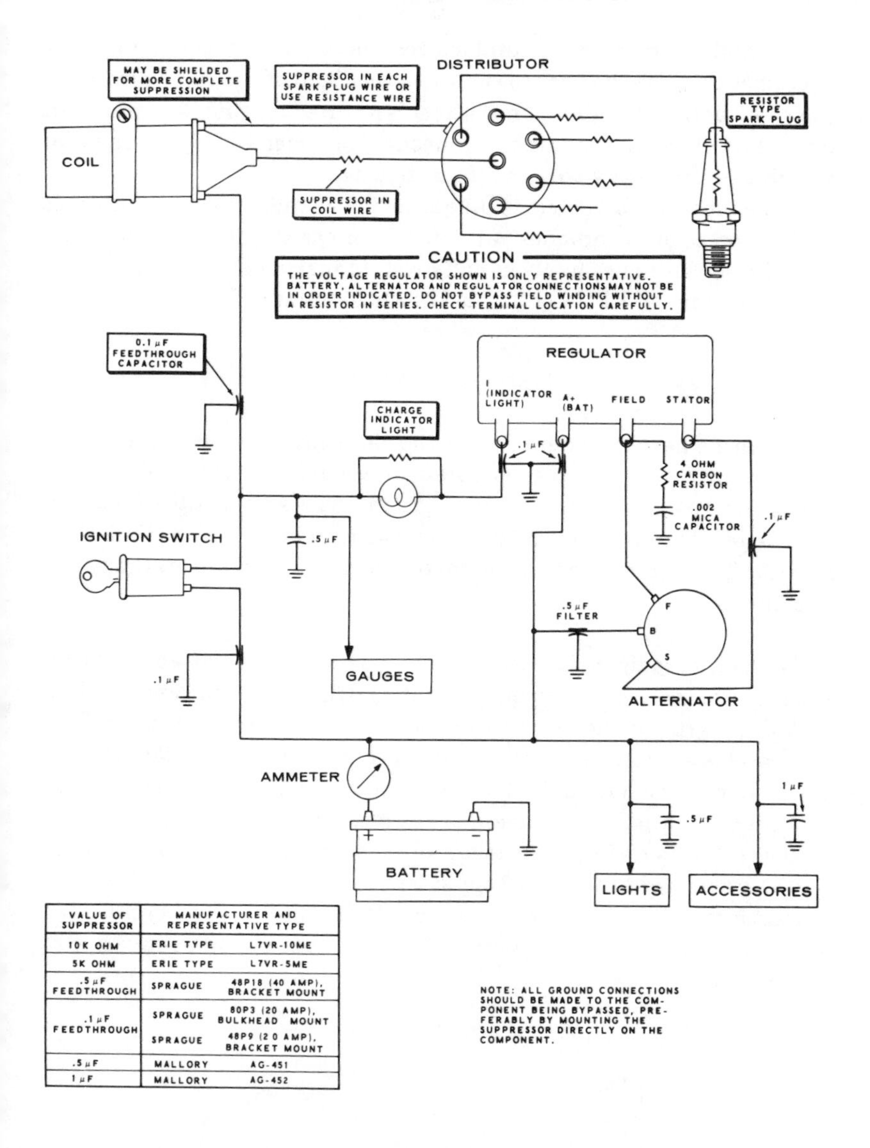

VALUE OF SUPPRESSOR	MANUFACTURER AND REPRESENTATIVE TYPE	
10K OHM	ERIE TYPE	L7VR-10ME
5K OHM	ERIE TYPE	L7VR-5ME
.5 μF FEEDTHROUGH	SPRAGUE	48P18 (40 AMP), BRACKET MOUNT
.1 μF FEEDTHROUGH	SPRAGUE	80P3 (20 AMP), BULKHEAD MOUNT
	SPRAGUE	48P9 (20 AMP), BRACKET MOUNT
.5 μF	MALLORY	AG-451
1 μF	MALLORY	AG-452

Fig. 13 Representation of RFI-proof vehicular electrical system.

(The drawings and charts in this chapter are courtesy of the Champion Spark Plug Company)

A final hint: many manufacturers of vehicular communication equipment advise the user to run the power lead of the radio directly to the "hot" battery terminal. The battery is an effective filter for most radio noise and it is best to make connection as close to its terminals as possible, using heavy wire to minimize loss.

Remember, each vehicle is unique as far as radio noise is concerned and suppression techniques can vary from car to car. Remember also that these techniques apply equally well to marine installations.

A summary of RFI-proofing is shown in Figure 13.

Marine Radio Noise

A marine gasoline engine poses the same problems as a land-based engine and the same noise suppression techniques are used. It is of utmost importance that a common ground bus is part of the electrical system. Most boats have a heavy copper ground bus that runs the length of the hull. The motor and one side of the electrical system are attached to this bus.

To reduce radio noise, the ground bus should be attached to an external ground area which makes contact with water. If an exposed keel is used, the ground bus may be attached with bonding cable to a keel bolt. If the keel is encased in fiberglass, a copper ground plate should be bolted to one side of the hull under the waterline and attached to the common bus. Grounding plates are available at most marine hardware stores.

Depth finding equipment is a prolific source of radio noise and the "hot" power lead to the unit should be bypassed with a 0.1 uF coaxial capacitor grounded to the frame of the unit. Bypassing the power lead of the radio equipment to ground with a coaxial capacitor is also recommended.

Chapter 13

RFI Roundup

Telephone Interference

Many CBers, amateurs and their neighbors have experienced telephone interference caused by transmitting equipment. The phone lines seem to act like an antenna, conducting the signal to the carbon microphone and compensating networks which serve to act as a detector, demodulating the signal and sending it to the earpiece.

In the early days, the complainant would contact the telephone company repair service when telephone interference arose. A repairman would come out and place a small capacitor across the telephone microphone, thus bypassing the signal rectification. This simple solution no longer works with today's complex telephones but many of the smaller, independent phone companies don't know it and much time and wasted effort has gone into solving the problem of telephone interference.

Modern telephones employ an equalizer network to reduce crosstalk and shape the audio response of the system for best voice transmission. Among other components, the network may have a pair of voltage-sensitive diodes that are rf sensitive. The problem is to keep the radio signal out of the network and the microphone.

A Simple Interference Eliminator

A simple interference filter that works with many telephones is shown in Figure 1. Illustration A shows two or more ferrite beads slipped over the telephone lines at the terminal strip. Up to three beads may be used. If interference persists, the filter circuit shown in illustration B can be tried. The components are installed in the telephone, not at the baseboard terminal block.

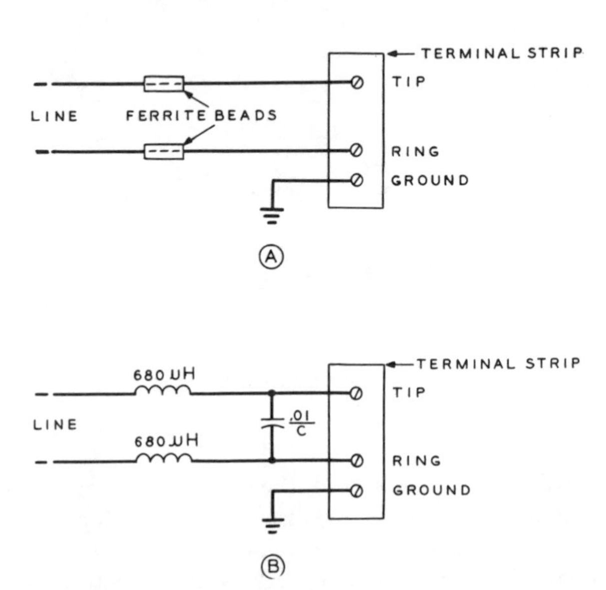

Fig. 1 Ferrite beads slipped over telephone line often reduce the RFI level in the earpiece. Suitable beads are Amidon FB-75-101 having a permeability of 5000. Small rf chokes can also be used.

The Bell System has suitable capacitors and inductors to do the job. The inductor is *part 1542A* and the capacitor is *part 40BA*. They should be installed at the connector block.

Useful information covering elimination of telephone RFI is contained in the *Bell System Practices* manual, Plant Series, Issue 6 (April, 1970), section 500-150-100. The manual also discusses interference problems caused by corroded connections, loose connections, abandoned drop wires still connected to the line and unterminated loops to other rooms in the home.

Modifications to the Telephone

In addition to the line filter, it is often necessary to place additional filtering inside the telephone. The compensation and volume limiting circuits must be protected from the rf signal.

The old style *model 300* series "French phones" are readily identified by a handpiece that is triangular in cross section and by an oval dial pedestal. Placing a .01 uF disc capacitor across the carbon microphone is usually effective in cleaning up remnants of interference.

The newer style *model 500* series "French phones" are identified by a handpiece that is rectangular in cross section and a bell loudness control on the underside of the dial pedestal. The pedestal is rectangular in shape. This model includes a network within the pedestal which is sensitive to rf. The network should be removed and replaced with a type 425J network that is impervious to rf energy. The network is furnished and installed by the telephone company. At the same time, the microphone should be bypassed as is done with the older phones.

The newest style *model 1550, 1600* and *1700* series "Touchtone" phones are similar in operation to the series 500 and require a 4010E replacement network. Additional bypass capacitors are required to complete the job and these are discussed in the *Bell System Practices* manual. The "Princess" series phones also require a complex bypass capacitor installation that is discussed in the manual.

In general, most telephone companies would rather remove the offending telephone and replace it with a modified telephone having the RFI suppression circuits installed than make an on-the-spot modification. The repair service or operations manager of your local telephone company can provide information as to the replacement phone sets.

Cardiac Pacemaker Interference

The life-saving cardiac pacemaker is a miniature electronic device implanted in the human body to regulate heart action. These devices *may* be susceptible to malfunction in a strong rf field of a nearby transmitter. The pacemaker provides heart-pacing pulses at a rate from 60 to 130 pulses per minute, delivering a stimulus when the heart rate drops below the preset pulse generator rate.

The owner of a pacemaker should consult the manufacturer to determine the rf level at which his unit may be affected before he subjects himself to a strong rf field. This applies particularly to engineers working at radar and transmitter sites, radio amateurs, and possibly to CB operators.

Military RFI Control Standards

Military agencies have extensive RFI control standards summarized in document MIL-STD-461B available through appropriate channels. This serves as a model and a guideline in developing industry specifications in accord with military requirements. The document is divided into many parts, separately bound, so an order should request MIL-STD-461B and all its parts. In addition, the document contains important lists of referenced publications covering military RFI specifications, electromagnetic compatibility handbooks, interference reduction and RFI testing procedures.

Electromagnetic Compatibility Courses

The in-depth problems of RFI and electromagnetic compatibility have grown so complex in recent years that a large portion of research and development work on modern commercial and military equipments is devoted to the measurement, analysis and control of RFI problems.

Courses that focus on these problems are now available in all large cities in the United States. Intended for practicing professionals who need a comprehensive, practical awareness of the RFI/electromagnetic compatibility field, such courses equip the participants with control techniques which they may apply to the design of their products. The courses last from one to five days.

Information on these programs may be obtained from The Institute of Electrical and Electronic Engineers, Inc., 345 East 47th St., New York, NY 10017 (phone: 212-644-7806).

Oscilloscope Pictures of RFI

Most electronic technicians, servicemen and engineers (as well as many radio amateurs) have an oscilloscope among their test instruments. This useful instrument may be used to provide clues to trace RFI in conjunction with an am receiver. The audio output of the receiver is fed into the vertical channel of the scope and the RFI can often be identified visually as a particular image on the screen.

The following pictures of various forms of RFI were recorded directly from a receiver output by a Sony TC-200 tape recorder. The tape deck was then connected to the vertical channel of a memory-type oscilloscope where the RFI image was held on the screen, thus allowing the pictures to be taken.

The scope is of little value in the identification of RFI when an SSB receiver is employed as the narrow bandwidth of reception and oscillator injection seem to make all types of noise appear similar on the scope. In the am mode, however, definite RFI characteristics can be seen.

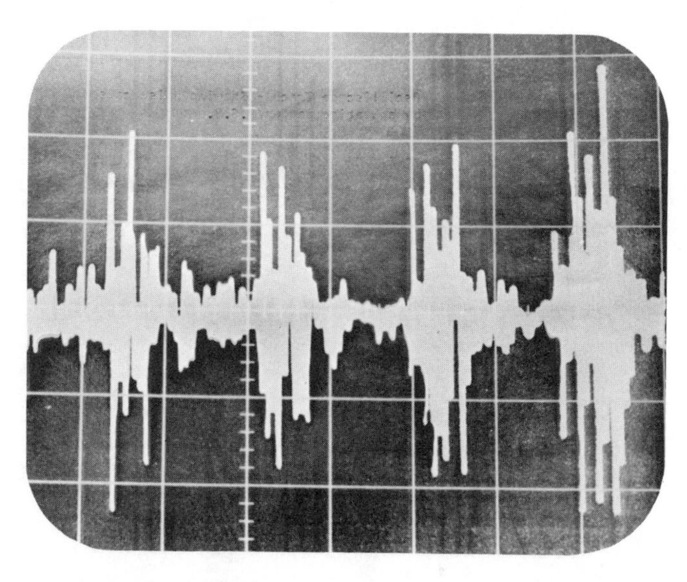

Fig. 1 Typical spark interference. This discharge was caused by a defective heating pad. Note the large amount of energy in each burst of interference.

Fig. 2 Electric fence interference is characterized by high levels of spark energy at a rapid repetition rate determined by the fence control box.

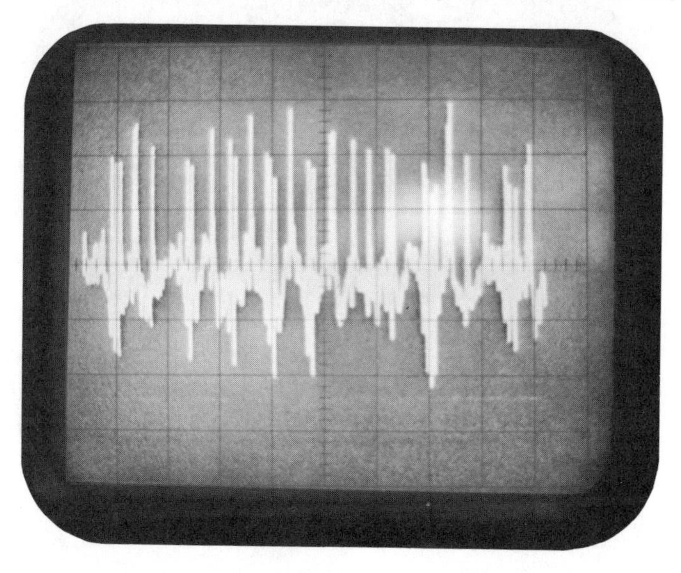

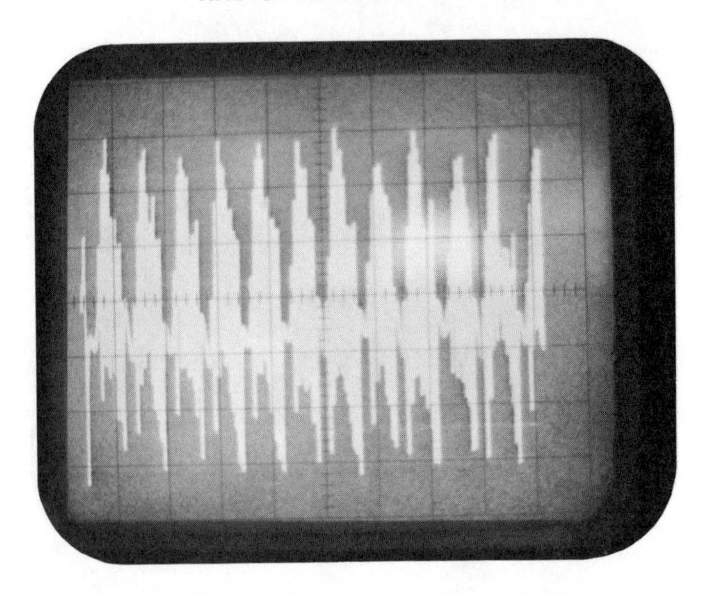

Fig. 3 High energy pulses of RFI from a Heliarc welder can cause severe disruption to radio and tv reception.

Fig. 4 Sharp RFI pulses from light dimmer cause havoc in nearby radios.

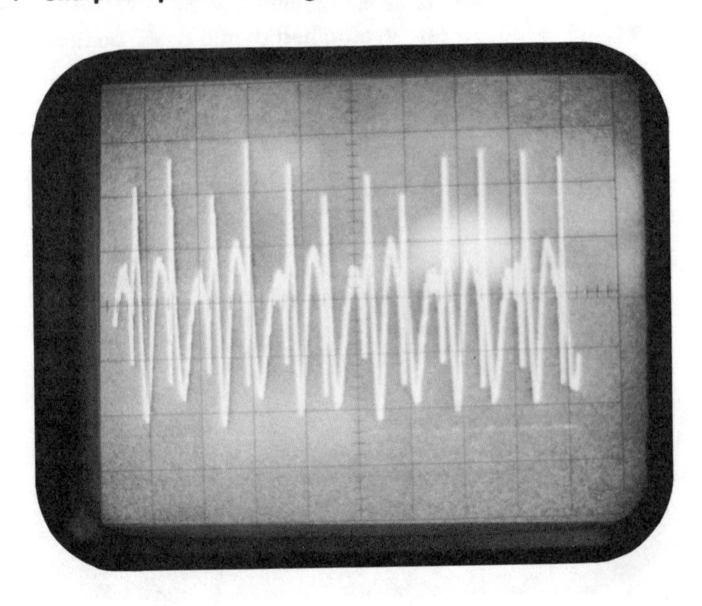

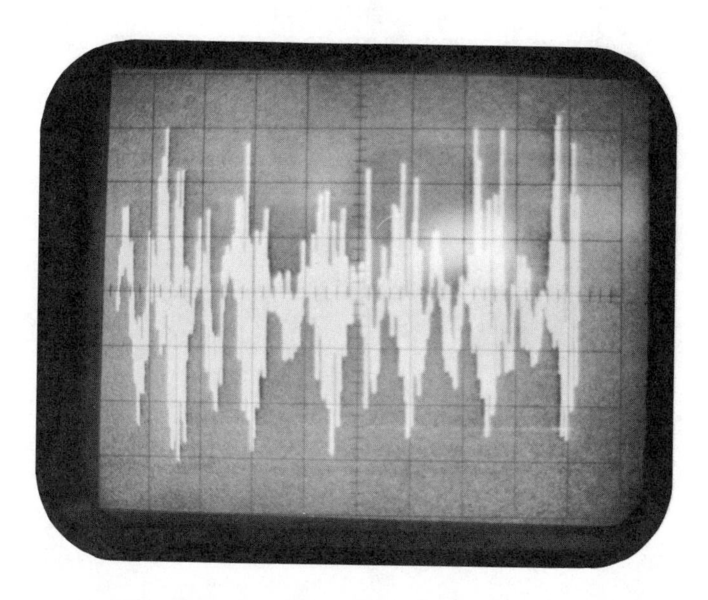

Fig.5 Brush-type motor commonly used in mixers, electric drills, razors and various power tools creates prolific RFI.

Fig. 6 Note the multiple pulse components in pattern of fluorescent light.

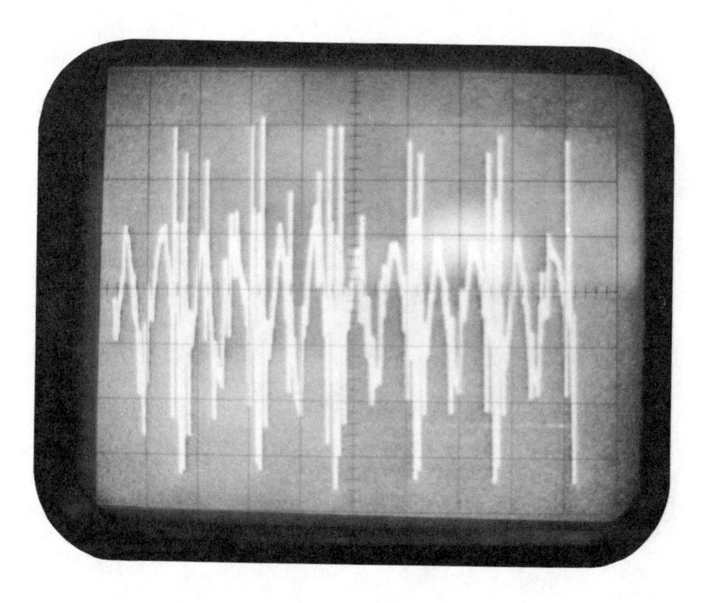

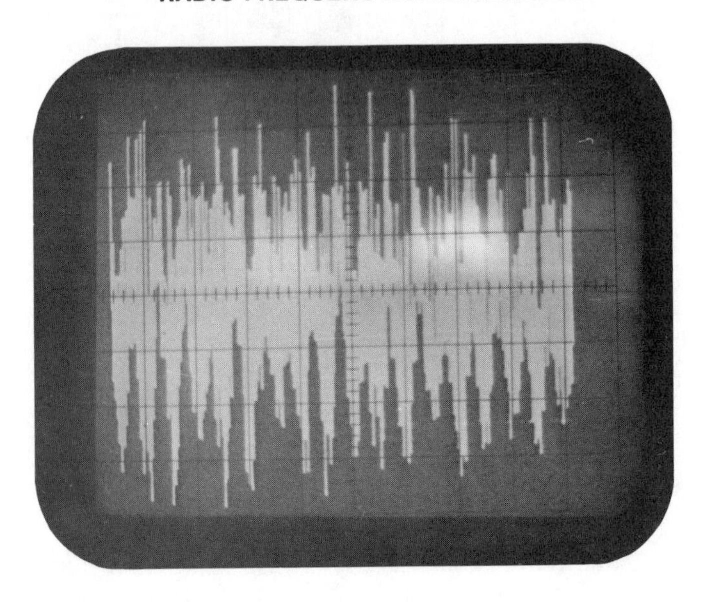

Fig. 7 Heavy corona discharge from power line has high energy content.

Fig. 8 Spark discharge from power line lightning arrester.

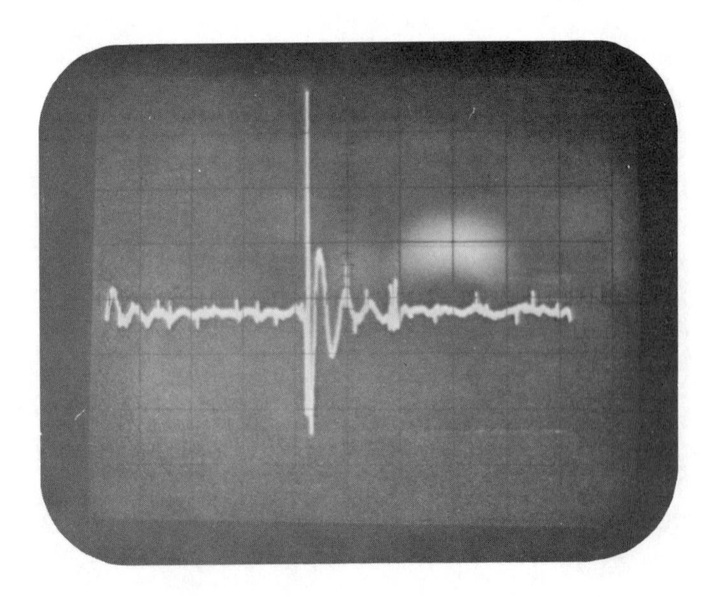

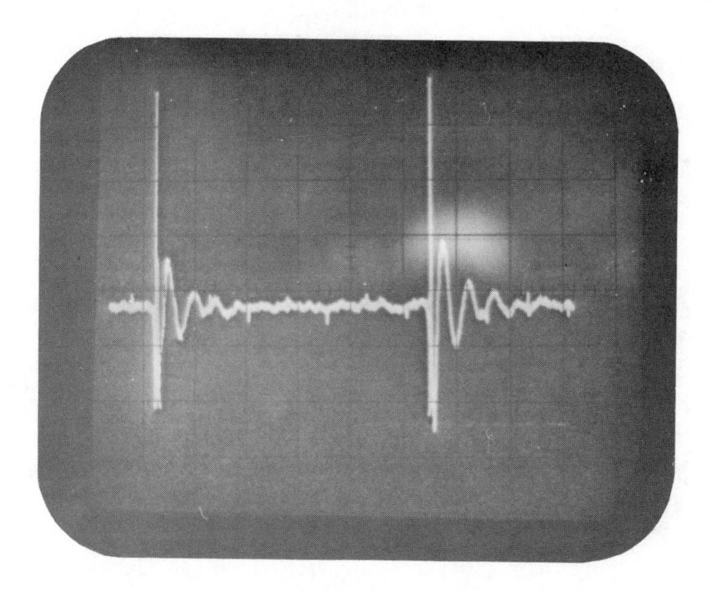

Fig. 9 Powerful pulses emitted by an unbonded pole-top switch.

Fig. 10 Loose hardware on power pole creates prolific RFI.

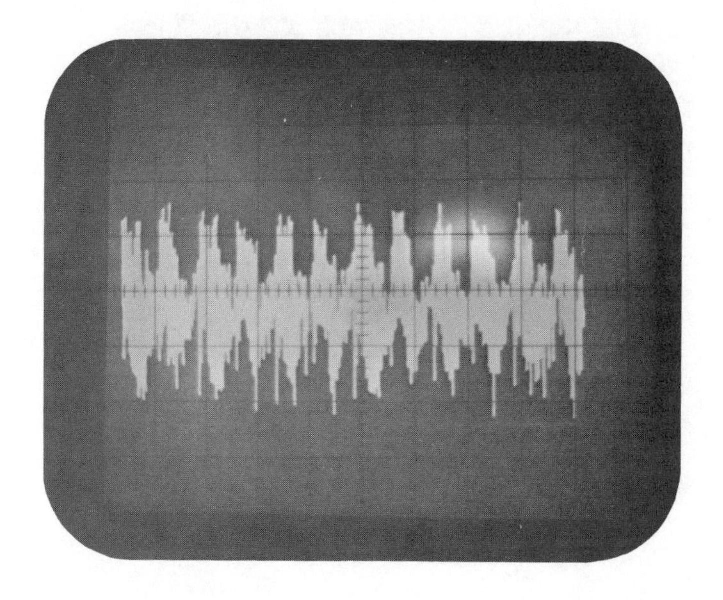

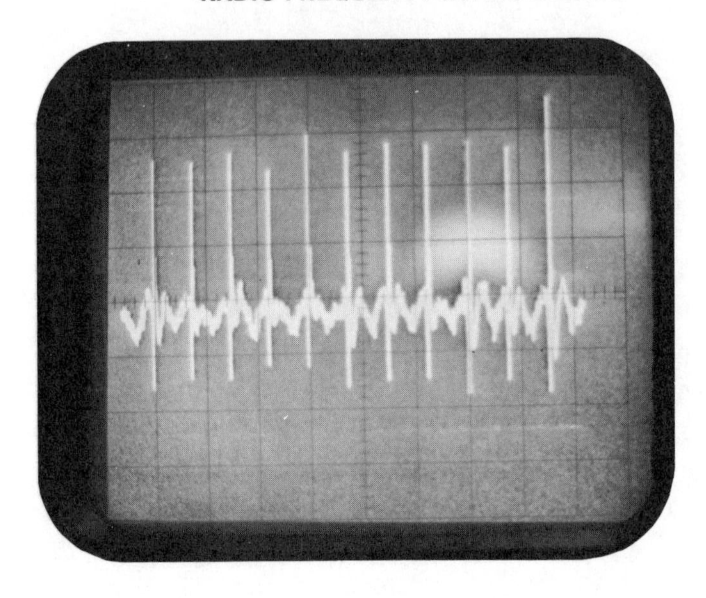

Fig. 11 Slack deadends on power line pole produce high, repetitive pulse.

Fig. 12 A loose connector on power line pole.

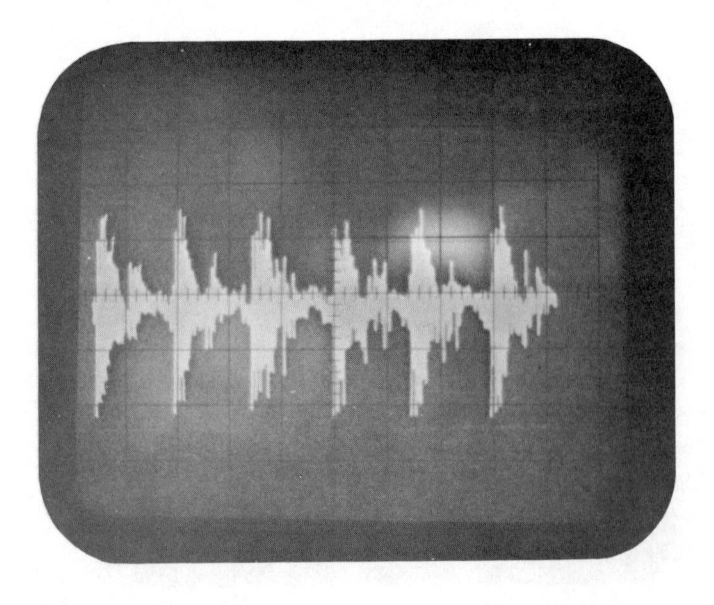

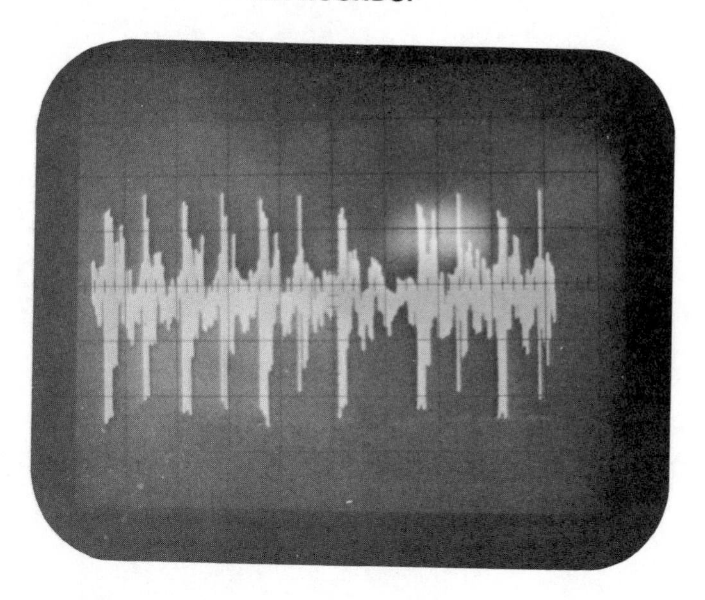

Fig. 13 Loose crossarm bond wire on pole.

Fig. 14 Ragged interference from a mercury vapor street light.

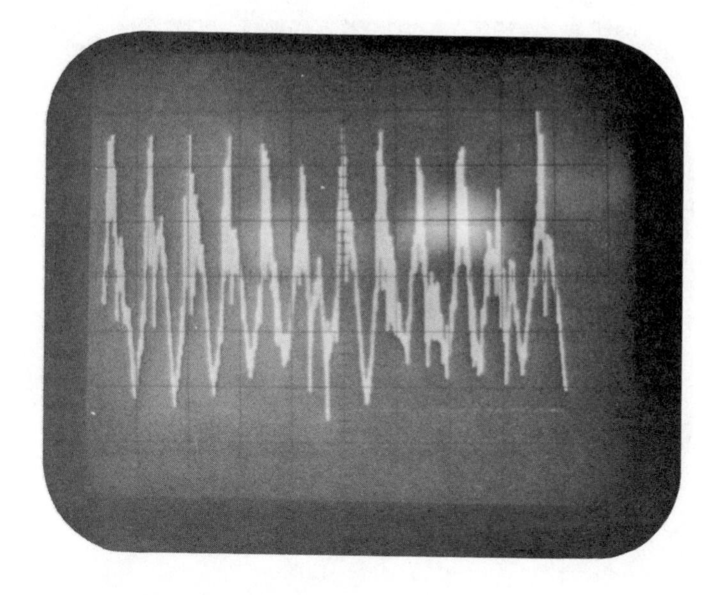

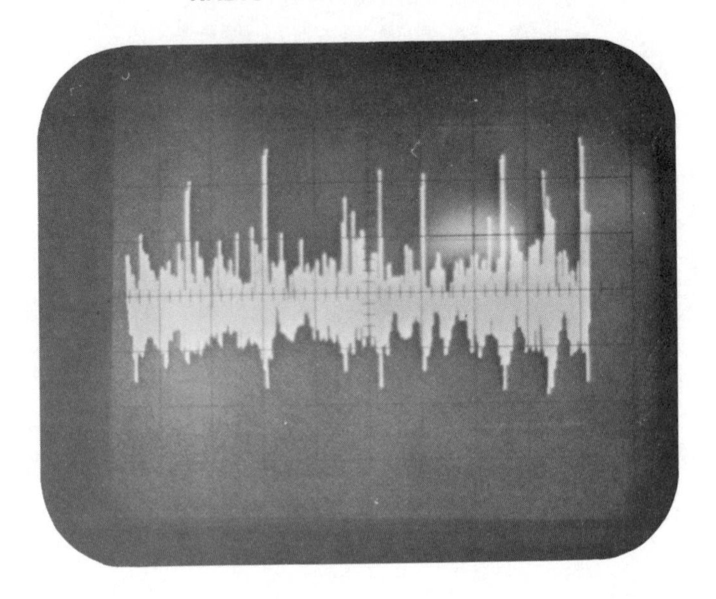

Fig. 15 RFI from the horizontal sweep oscillator of a nearby tv receiver.

Fig. 16 Note the complexity of a typical tv signal as seen on the oscilloscope.

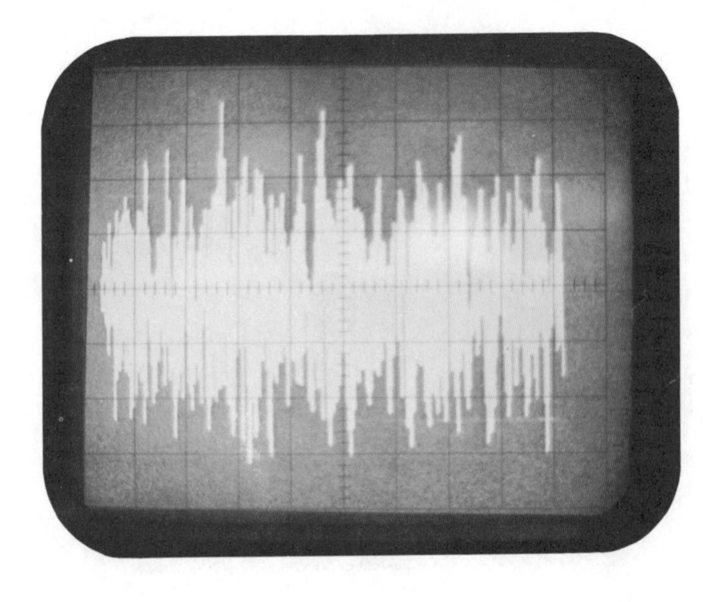

These photographs were made with the cooperation of the Electronics Department, Chaffey Community College, Cucamonga, California.

Smoke Detector RFI

Some smoke detectors, particularly the *optical refraction sensor* designs, can be triggered by a strong radio signal, setting off the alarm. Little information is available from the manufacturers of these devices and the only effective measure found to date to suppress RFI is to open the unit and bypass the center movable arm of the sensitivity potentiometer to the outer arms with .001 uF disc ceramic capacitors. The original setting of the control should not be changed.

The *ionization sensor* design has a radioactive ionization source and if the container is broken the radioactive chemical could be spread around. It is strongly recommended that these devices not be opened or modified.

It should be remembered that any modification to a smoke detector *may* prevent it from performing as intended. On the other hand, the device may not perform in the presence of an rf field if unmodified.

It is recommended that the owner of a smoke detector troubled by RFI contact the American Radio Relay League, 225 Main St., Newington, CT 06111 for the latest information on this troublesome problem before modifications are made to any device.

RFI Assistance List

The following material was compiled by Harold Richman, W4CIZ, a member of the ARRL RFI Task Group. This list summarizes the names, addresses and phone numbers of prominent home entertainment equipment manufacturers who provide information and assistance to owners of their products that experience interference. If you have an RFI or TVI problem, contact the manufacturer concerning warranty, service information, and instructions. Give all details of the problem, along with model type, serial number, date of purchase and other pertinent information.

Thanks to the ARRL for their permission to reprint this material and to W4CIZ for compiling this useful list.

Admiral

No longer in business. For parts, tel. 800-447-8361.

Akai America

Akai products include audio tape recorders, video tape recorders, a-m/fm receivers, speaker systems and related accessory products. Inquiries related to RFI should be addressed to the Customer Service Department, 800 W. Artesia Blvd., Compton, CA 90220, or to P. O. Box 6010, Compton, CA 90224, tel. 213-537-3880. "Upon receipt of these inquiries, we will investigate the situation and, to our utmost, try to resolve the customer's problems."

Allen Organ Company

When a complaint is received via the dealer, Allen Organ Co. sends the dealer an informational service bulletin on RFI and sufficient components to cover all amplifiers in the affected instrument. This service is offered at no cost to the customer. Refer RFI problems to the local Allen dealer. Inquiries may be made to Mr. David L. George, National Service Manager, Macungie, PA 18062, tel. 215-966-2200.

Altec Lansing International

Customer RFI problems are referred to the authorized Altec warranty stations located nationwide and denoted by an information card furnished with each piece of equipment. Unusual situations are, at the option of the warranty station, referred to Altec Customer Service, 1515 W. Katella Ave., Anaheim, CA 92803, tel. 714-774-2900, or to the Engineering Department, 1515 S. Manchester Ave., Anaheim, CA 92803, Attention: Chief Engineer, Electronics.

Apple Computer, Inc.

"Our products include business, professional, educational, scientific, industrial and home computers, peripheral devices, and software. These products are designed to be compliant with the FCC guidelines covering Class A and Class B computer devices. Inquiries related to RFI should be addressed to any of our more than 800 dealer-operated Level One service centers. If the service technicians there are unable to solve the situation, they will contact our Corporate Engineering Services Group."

Arvin Industries, Inc., Consumer Electronics Division

Customer problems involving RFI should be referred to Mr. John Currey, Manager Engineering Support Group, E. 15th St., Columbus, IN 47201, tel. 812-372-7271.

Audio Research Corporation

In the event of an RFI problem, the customer may write to Mr. Richard Larson, Chief Engineer, 6801 Shingle Creek Pkwy., Minneapolis, MN 55430, tel. 612-566-7570.

Baldwin Piano and Organ Company

"RFI complaints are usually handled by the local Baldwin service technician. Factory personnel are available to assist a technician when needed. Baldwin maintains its own staff of technical representatives who travel in the field and may be called upon to assist a dealer technician with difficult problems, including RFI. Several Baldwin Technical Manual Supplements are available with specific instructions for RFI suppression on specific models. This information is readily available upon request. Inquiries may be directed to Mr. Gilbert C. Carney, Manager Organ Technical Service, Baldwin Piano and Organ Co., 1801 Gilbert Ave., Cincinnati, OH 45202, tel. 513-852-7838."

Bogen Division of Lear Siegler, Inc.

"Bogen Division manufactures professional, commercial and industrial sound equipment. In the event of an RFI problem with any Bogen unit, write for the division's free Field Service Bulletin No. 59 about RFI signal interference, or contact Allen Guthman, Service Manager, Bogen Division/LSI, Box 500, Paramus, NJ 07652, tel. 201-343-5700."

Carver Corporation, Inc.

Carver Corporation manufacturers high-fidelity components. "Problems pertaining to RFI should be directed to our service manager, Mr. Philip Fenner, P. O. Box 664, 14304 N.E. 193rd Pl., Woodinville, WA 98072, tel. 206-487-3483."

Conn Keyboards, Inc.

RFI complaints should be referred to the local Conn dealer, whether instrument is in or out of warranty. Factory assistance is available to the dealers who are unable to correct the RFI. RFI problems encountered within the term of instrument warranty are usually corrected by the selling dealer without cost to the organ owner. Contact Mr. Thomas A. Umbaugh, National Service Manager, 350 Randy Rd., Carol Stream, IL 60187, tel. 312-653-4330.

Crown International

"Crown International is the manufacturer of high-end audio products. RFI suppression is incorporated in the design of the product. If a customer should encounter an RFI problem, he may contact the Technical Services Department of Crown International, 1718 W. Mishawaka Rd., Elkhart, IN 46517."

Curtis Mathes

Curtis Mathes products include color TVs and stereos (100% solid state) in portable, console and combination configurations. Customer complaints involving RFI should first be resolved at the retail-dealer level. If not satisfied, then the complaint should be made in writing to the Consumer Relations Department giving all details of the problem, along with the model information, serial number, date of sale, dealer and service history. Each complaint will be handled individually. Write to Curtis Mathes Manufacturing Co., Curtis Mathes Pkwy., Athens, TX 75751, tel. 800-527-7646, Texas only tel. 800-492-9543.

Delco Electronics, Division of GM Corporation (see GM Corp.)

Dumont (see Emerson Quiet/Kool Corp.)

Electra Company, Division of Masco Corporation of Indiana

Electra Co. asks that RFI problems with "Bearcat," its automatic scanning radio, be referred to its service department at 300 E. County Line Rd., Cumberland, IN 46229, tel. 317-844-1440.

Emerson Quiet/Kool Company

Mr. Jerome Roth reports that his company has not made TVs or audio devices since 1972. As a continuing gesture of goodwill, however, Mr. Roth suggests that customers may refer RFI problems with equipment previously marketed by Emerson Quiet/Kool Co. to him for recommendations, at the mailing address below. *Do not confuse* this company with Emerson Radio Corp., which is an entirely different publicly owned corporation. Contact Emerson Quiet/Kool Co., P. O. Box 300, Woodbridge, NJ 07095, tel. 201-381-7000.

Emerson Radio Corporation

Customers may refer RFI inquiries related to Emerson Radio Corp. TV and radio problems to Mr. Dave Buda. Emerson Radio does not supply filters. The new address is: Emerson Radio Corp., One Emerson Way, Secaucus, NJ 07094, tel. 201-865-4343.

Epicure Products, formerly Elpa Marketing Industries, Inc.

"Complaints are handled with respect to parts and labor on an individual basis. Necessary modifications for RFI are made on a no-charge basis for parts and labor during the term of instrument warranty. Beyond warranty, modification parts are available free of charge. The customer then pays for labor involved in the installation of the parts. Refer RFI problems to Mr. John F. King, National Service Manager, 25 Hale St., Newburyport, MA 01950, tel. 800-225-7932."

Fisher Corporation

Fisher Corporation asks that RFI problems involving a Fisher product be handled as follows: request assistance from the local selling dealer or request assistance from the local Fisher authorized service station (a list is packed with every Fisher unit). Contact with local Fisher agencies is the preferred method of handling. Fisher's service coordination group maintains close communications with Fisher authorized service stations and Fisher's Engineering Department, and works under the supervision of the office of the National Service Manager. If the problem cannot be solved at the first two service levels, contact Service Coordination, 21314 Lassen St., Chatsworth, CA 91311, tel. 213-998-7322.

Garrard/Plessey Consumer Products

Garrard advises the consumer on methods that may eliminate RFI. In unusual cases where the suggestions are ineffectual, customers should refer the RFI problem to Mr. Al Pranckevicus, National Service Manager, 85 Sherwood Ave., Farmingdale, NY 11735, tel. 516-293-2400.

General Electric Company

RFI problems involving G.E. television receivers should be referred to the nearest General Electric Customer Care Service Operation. If G. E. Customer Care Service is unable to correct the RFI, the customer should refer the problem to General Electric Co., Mr. J. F. Hopwood, Manager of Consumer Affairs, Appliance Park, Louisville, KY 40225, tel. 502-452-3754. All RFI problems involving G. E. radios, record players and other audio products should be referred to Manager of Consumer Counseling, Mrs. Patricia C. Cleary, Electronics Park, Bldg. 5, Syracuse, NY 13221, tel. 315-456-3388.

General Motors Corporation

"From time to time you may have questions concerning the electromagnetic compatibility of mobile transmitters when installed on General Motors vehicles. To help avoid such questions from arising, it is urged that care be taken to follow any applicable GM service procedures. The local GM Service Manager for the Car or Truck Division whose vehicle is involved should be contacted for information about such service procedures. If you are unable to obtain such assistance locally or if questions nevertheless arise, we have established a central contact point for all such inquiries. Accordingly, you should direct your inquiries to: Mr. Henry J. Lambertz, GM Service Research (GMSR), Service Development Center, 30501 Van Dyke, Warren, MI 48090, tel. 313-492-8448. He will direct your inquiries to the appropriate divisions or staff within GM and follow up to see that appropriate action is taken."

Gulbransen, Division of CBS Musical Instruments, Inc.

Gulbransen cooperates with dealers and customers in offering suggested solutions to RFI. Gulbransen does not reimburse the consumer for servicing. When extreme cases are encountered because of the proximity of the transmitter and relative power, however, the dealer may sometimes absorb the cost of servicing RFI problems. Customers should refer RFI problems to the local dealer. Inquiries may be directed to Mr. J. A. Iacono, Consumer Service Supervisor, 100 Wilmot Rd., Deerfield, IL 60015, tel. 800-323-1814.

Hammond Organ Company

"RFI difficulties are usually handled by the local Hammond dealer service technician. Hammond maintains a staff of technical service representatives who travel in the field and may be called upon to assist local dealer technicians with difficult or unusual service problems, including RFI." Hammond states that the services of the Engineering and Technical Field Service Departments under its control are provided to consumer and dealer without charge. RFI problems should be referred to the local Hammond dealer. Inquiries may be directed to the Hammond Technical Service Department, 4200 W. Diversey Ave., Chicago, IL 60639, Attention: Jerry J. Welch.

Harman/Kardon, Inc.

RFI problems should be directed to Harman/Kardon at 240 Crossways Park West, Woodbury, NY 11797, tel. 516-496-3406, Attention: Customer Relations Dept.

Heath Company

Heath Co. suggests that, for fastest service on matters related to RFI regardless of the product line involved, customers may now reach the Technical Consultation Department by either writing directly to that department at Heath Co., Benton Harbor, MI 49022, or by using a new direct-line telephone system to the department by calling 616-982-3302. Do not write to an individual.

Hitachi Sales Corporation of America

"Our primary products are TVs, radios, tape recorders, hi-fi components and video tape recorders. Hitachi Sales Corp. of America attempts to cure each RFI problem on an individual basis. Customers should provide model number and information concerning the nature of the problem. RFI problems should be referred to the nearest Hitachi Regional Office." *Eastern Regional Office,* 1200 Wall St. West, Lyndhurst, NJ 07011, tel. 201-935-8980, Attention: Service Dept. *Mid-Western Regional Office,* 1400 Morse Ave., Elk Grove Village, IL 60007, tel. 312-593-1550, Attention: Service Dept. *Western Regional Office,* 612 Walnut, Compton, CA 90220, tel. 213-537-8383, Attention: Service Dept. *Southern Regional Office,* 510 Plaza Dr., College Park, GA 30349, tel. 404-763-0360, Attention: Service Dept.

J. C. Penney Company, Inc.

J. C. Penney Company asks that customers with RFI problems contact their nearest J. C. Penney store for personal assistance. J. C. Penney Company, Inc., 1301 Avenue of the Americas, New York, NY 10019.

Kenwood Electronics, Inc.

Kenwood asks that customers with RFI problems take the affected unit to an authorized service center where an adjustment will be made at no cost to the customer if the product is properly registered with Kenwood and is within warranty. It is suggested that prior authorization for the return be obtained from Mr. Toshi Furutsuki, 1315 E. Watsoncenter Rd., Carson, CA 90745, tel. 213-518-1700.

Lafayette Radio Electronics Corporation

"Customers should refer RFI problems involving Lafayette products to the local dealer. If the dealer cannot alleviate the problem, the customer may contact Mr. Charles Tanner, Vice President Administration, 111 Jericho Tpk., Syosset, NY 11791, tel. 516-921-7700."

Lowrey Division of Norlin Music, Inc.

"Lowrey customers should refer RFI problems to the local Lowrey dealer or certified Lowrey technician. Lowrey provides all technicians with technical literature regarding RFI and will provide assistance to local service organizations through its staff of field technical representatives when needed. Inquiries may be directed to Mr. Larry R. Thomas, Director of Product Service, 707 Lake Cook Rd., Deerfield, IL 60015."

Magnavox Consumer Electronics Company

"RFI problems are usually handled by the local Magnavox Authorized Service Center. Technical assistance in resolving such problems is provided by the Magnavox Field Service Staff through four Area Service Offices. Technicians or customers may refer unusual RFI problems involving Magnavox products to their nearest Area Service Center." In the *New York area* contact Magnavox Consumer Electronics Co., 161 E. Union Ave., East Rutherford, NJ 07073. In the *Chicago area* contact Magnavox Consumer Electronics Co., 7510 Frontage Rd., Skokie, IL 60077. In the *Atlanta area* contact Magnavox Consumer Electronics Co., 1898 Leland Dr., Marietta, GA 30067. In the *Los Angeles area* contact Magnavox Consumer Electronics Co., 2645 Maricopa St., Torrance, CA 90503.

Marantz (see Superscope)

McIntosh Laboratory, Inc.

"McIntosh has a number of authorized service agencies located throughout the country. Customers will be assisted to receive prompt help. RFI and other service-related problems can be directed to Mr. John Behory, Customer Service Manager, 2 Chambers St., Binghamton, NY 13903, tel. 607-723-3512."

MGA Mitsubishi Electric Sales America, Inc.

MGA is the new sales and service representative for the Mitsubishi Electric Corp. RFI reports from the field, beyond the dealer's capability to resolve and in which MGA becomes involved, are handled on an individual basis, as in the past. "All attempts will be made to give customer satisfaction." MGA suggests that requests for assistance be addressed to 3030 E. Victoria St., Compton, CA 90221, or the Service Department may be contacted by telephone, toll free, at 800-421-1132. Mr. Ken Kratka is the new National Service Manager.

Midland International Corporation

Midland policy remains the same. If any RFI problems are encountered with Midland portable black-and-white and color TVs or audio and radio products, individuals should contact Mr. Dennis Oyer, Vice President Customer Service, P. O. Box 1903, Kansas City, MO 64141, or at 1690 N. Topping, Kansas City, MO 64120, tel. 816-241-8500.

Montgomery Ward

Service for RFI should be obtained from the nearest Montgomery Ward location. If service is not obtainable locally, the customer may write to: Customer Service Product Manager, Corporate Offices, Montgomery Ward Plaza 4-N, Chicago, Il 60671. The Montgomery Ward field service organization can call upon factory and corporate engineering talent for assistance in handling difficult RFI problems.

Morse Electro Products Corporation

"RFI complaints related to Morse entertainment products may be referred to Mr. Phillip Ferrara, Service and Parts Dept., 3444 Morse Dr., Dallas, TX 75221, tel. 214-337-4711 or 800-527-6422."

Nikko Audio

"Nikko's line of products includes stereo receivers, tuners, amplifiers, combination preamp and main-amp pairs, tape decks and signal processors. For information and assistance with any Nikko products, inquiries should be made to Mr. Robert Fontana, National Service Manager, Service Dept., 320 Oser Ave., Hauppauge, NY 11787, tel. 516-231-8181."

North American Phillips Corporation

This corporation no longer manufactures its own RFI-prone products. (See Sylvania.)

Nutone Division

"Refer RFI problems to Mr. Norman W. Aims, Field Service, Scovil Housing Products Group, Madison and Red Bank Rds., Cincinnati, OH 45227, tel. 513-527-5415."

Panasonic Company

When instances of RFI occur, the customer should contact Panasonic at the following address: Panasonic Co., Division of Matsushita Electric Corp. of America, One Panasonic Way, Secaucus, NJ 07094, Attention: Supervisor of Quality Assurance Group, tel. 201-348-7000. The customer should provide model number, serial number and information concerning the problem. Upon review of the problem, the customer will be contacted and advised where to return the unit for corrective repair. "Panasonic will absorb both parts and labor costs in these instances."

Phase Linear Corporation

"RFI problems should be directed to Phase Linear Service Dept., Rick Bernard, Service Manager, 20121 48th Ave. West, Lynnwood, WA 98036, tel. 206-774-8848. In-house articles regarding RFI cures are available upon request at no charge."

Quasar Company (Matsushita Corporation of America)

For a high-pass filter, the consumer should contact Quasar Co., Consumer Relations Manager, Mr. George Datillo, 9401 W. Grand Ave., Franklin Park, IL 60131, tel. 312-451-1200. Model and serial number of the receiver and frequency of the interfering signal, if known, should be included with the written request, as well as whether sound or picture or both are affected. The Quasar distributor serving the local area should be contacted relative to any other interference problem that is unique to Quasar products.

Radio Shack

"Customers who encounter unique interference problems involving Radio Shack audio products may write to Mr. Dave Garner or Mr. Al Zuckerman, Product Development Engineers, National Headquarters, 1100 One Tandy Center, Fort Worth, TX 76102, tel. 817-390-3205."

RCA Consumer Electronics

"RFI problems involving both TV and audio products may be referred to Mr. J. J. Sanchez, 600 N. Sherman Dr., Indianapolis, IN 46201, tel. 317-267-6448. Requests for filters should include model number and serial number of the RCA television receiver. Filter installation charges will be the customer's responsibility."

Rodgers Organ Company, Division of CBS Musical Instruments, Inc.

RFI problems involving the Rodgers Organ may be referred to Custom Organ Test Department, 1300 N. East 25th Ave., Hillsboro, OR 97223, tel. 503-648-4181.

Rotel of America, Inc.

Stereo receivers, amplifiers, tuners and tape decks are made by Rotel. RFI problems should be referred to Michael Gregory, National Service Manager, 13528 S. Normandie Ave., Gardenia, CA 90249. "RFI problems will be handled according to the terms of our limited warranty."

Sansui Electronics Corporation

"RFI problems should now be directed to Mr. Frank Barth, Vice President Frank Barth, Inc., 500 5th Ave., New York, NY 10110, tel. 212-398-0820. Frank Barth, Inc. is the new advertising and public relations agency representing Sansui. Mr. Barth will direct the customer to an appropriate Sansui Service Center." A Sansui representative has previously stated that all Sansui products are carefully checked prior to final engineering commitments for susceptibility to RFI. "Units are often taken to high-rf-level areas such as New York City to determine any design flaws."

Sanyo Electric, Inc.

"In the event an RFI problem should occur, the customer is requested to take the set to the nearest Sanyo authorized repair station. Transportation to and from the shop is the responsibility of the customer. Should the shop not alleviate the problem, either the customer or the shop should contact Mr. Brad Coulter, Consumer Relations Manager, Sanyo Electric, Inc., Electronics Division, 1200 W. Artesia Blvd., Compton, CA 90220, tel. 213-537-5830."

Scientific Audio Electronics, Inc.

"Refer RFI inquiries to Mr. Michael L. Joseph, National Marketing Manager, or contact Mr. Robert Hunt, National Service Manager, 701 E. Macy St., Los Angeles, CA 90012, tel. 213-489-7600."

H. H. Scott, Inc.

This manufacturer offers a simple instruction sheet to aid customers in resolving problems involving rf pickup. The information includes suggestions about suitable equipment grounding, power-line bypassing and hints and suggestions on how to determine where rf is entering the equipment. "Customers should refer any RFI problems to Mr. D. F. Merryman, Engineering Dept., 20 Commerce Way, Woburn, MA 01801, tel. 617-933-8800."

Sears, Roebuck and Company

Sears asks that customers with an RFI problem involving a Sears product contact the nearest Sears service department for assistance. Inquiries may be directed to Mr. R. C. Good, Manager Marketing Communications, Home Appliances, Dept. 703, Sears Tower, Chicago, IL 60684, tel. 312-875-8366.

Sharp Electronics Corporation

"Sharp Electronics will, with proof of purchase, supply customers with a Drake TV-300 high-pass filter at no cost. Audio rectification problems are handled on an individual basis by the Service Department. Refer RFI problems to Service Manager, 2 Keystone Pl., Paramus, NJ 07652, tel. 201-262-9000."

Sherwood, Division of Inkel Corporation

Customers with interference problems should contact Mr. David Daniels, Vice President Marketing, 17107 Kingsview Ave., Carson, CA 90746, tel. 213-515-6866.

Shure Brothers, Inc.

The manufacturer recommends the use of balanced-line, low-impedance microphones and cables. If an RFI problem persists after the above measures have been taken, the customer should contact Shure Brothers, Inc. with specifics so that they may be able to help solve the problem. Refer RFI problems to Customer Services Dept., 222 Hartrey Ave., Evanston, IL 60204, tel. 312-866-2553.

Sony Corporation of America

"Our primary products are color television, black-and-white television, video tape recorders, stereo equipment, audio components and word-processing equipment. RFI assistance is provided through regional service managers of Sony Factory Service Centers through the Customer Care Dept. An RFI booklet is available from the company on request. Sony Corp., 47-47 Van Dam St., Long Island City, NY 11101, tel. 212-361-8600."

Sound Concepts

"We handle all RFI complaints at our main laboratories at 27 Newell Rd., Brookline, MA 02146, tel. 617-566-0110. We request that the offending unit be accompanied by a description of the nature of the RFI; there is no charge for this service."

Soundesign Corporation

"Soundesign Corp./Acoustic Dynamics requests that all service problems relating to nonstereo merchandise be referred to Mr. Thomas R. Greene, Administrative Vice President, 34 Exchange Pl., Jersey City, NJ 07302, tel. 201-434-1050. All service problems on stereo merchandise are to be referred to our authorized service centers. The nearest one can be found by calling toll free in the continental U.S., 800-631-3092."

Superscope/Marantz Corporation

Superscope/Marantz manufacturers a-m/fm receivers, tuners, amplifiers, tape recorders, record players and audio systems. In the event of special RFI cases resulting from extremely high fields, contact the Technical Services Dept. at Superscope corporate offices. "Modifications necessary to resolve such RFI problems are provided to customers on an individual basis." Superscope/Marantz Corp., 20525 Nordhoff St., Chatsworth, CA 91311, tel. 213-998-9333. For Service Dept., call toll free, 800-423-5224, Attention: Mr. Albert Almeida, Technical Service Manager.

Sylvania/Philco, Division of North American Phillips Corporation

Sylvania policy remains as follows: "Factory field service and field engineering personnel work together to solve many of the TVI and audio rectification problems. If the consumer has an interference condition, he should contact his local dealer. He is in touch with the manufacturer's services that will help resolve it." Consumers should contact the dealer and work through his services first. RFI problems are handled on an individual basis. Sylvania has available for their technicians an excellent pictorial TVI training manual titled, *Diagnosis, Identification and Elimination of TVI.* Sylvania/Philco, Mr. Jack Berquest, Manager Service Training, Consumer Electronics Division, 700 Ellicott St., Batavia, NY 14020, tel. 716-344-5000.

Tandberg of America, Inc.

When RFI occurs in Tandberg products, the manufacturer suggests that the unit be returned to them. "We will do any modification possible to eliminate the RFI." Authorization should be obtained from Mr. Tor Sivertsen prior to return of the unit. Mr. Tor Sivertsen, Technical Vice President, Labriola Ct., Armonk, NY 10504, tel. 914-273-9150.

Thomas International Electronic Organs, Division of Whirlpool Corporation

"RFI is usually resolved at the dealer level. If the manufacturer's field service is made aware of a consumer complaint regarding RFI, they contact the seller and advise him on how to eliminate the problem." Thomas has six field service engineers. In the event of a call for assistance, an engineer personally contacts the consumer by telephone and makes an appointment to visit the home of the consumer to correct the RFI condition, with or without the dealer's technician. "We do not charge the consumer for this service." Refer RFI complaints to the dealer. Inquiries may be directed to Mr. Daniel E. Hofer, Manager Field Service, 7300 Lehigh Ave., Chicago, IL 60648, tel. 312-647-8700 or 800-323-4301.

Toshiba America, Inc.

Customers should contact the nearest regional office, an updated listing of which appears below, for obtaining assistance in solving RFI problems involving Toshiba televisions, radios, tape products, amplifiers, tuners and receivers. Mr. Stanley Friedman, National Service Manager, 82 Totowa Rd., Wayne, NJ 07470, tel. 201-628-8000. Mr. Sy Rosenthal, *Eastern Regional* Service Manager, 82 Totowa Rd., Wayne, NJ 07470, tel. 201-628-8000. Mr. Ray Holich, *Mid-West Regional* Service Manager, 2900 MacArthur Blvd., Northbrook, IL 60062, tel. 312-564-5110. Mr. C. B.

Monroe, *Southwest Regional* Service Manager, 3300 Royalty Row, Irving, TX 75062, tel. 214-438-5814. Mr. S. Ito, *Western Regional* Service Manager, 19515 S. Vermont Ave., Torrance, CA 90502, tel. 213-538-9960.

U.S. JVC Corporation

"Inquiries related to RFI involving JVC products may be referred to Mr. T. Sadato, Chief Engineer, 41 Slater Dr., Elmwood, NJ 07407, tel. 800-526-5308."

U.S. Pioneer Electronics Corporation

"Contact: Mr. Andrew Adler, *Eastern Region,* 75 Oxford Dr., Moonachie, NJ 07074; Mr. John Noa, *Southern Region,* 1875 Walnut Hill Ln., Irving TX 75062; Mr. Clarence Skroch, *Western Region,* 4880 W. Rosecrans Ave., Hawthorne, CA 90250; Mr. Daniel Brostoff, *Mid-West Region,* 737 Fargo Ave., Elk Grove Village, IL 60007."

Wells-Gardner Electronics Corporation

"Wells-Gardner is a private-label manufacturer of consumer products. Inquiries related to RFI should be referred to our private-label customers whose address appears on the model-number label attached to the product. Special problems which may be encountered by private-label customers are usually referred to Wells-Gardner, Mr. Harry McComb, Service Manager, 2701 N. Kildare Ave., Chicago, IL 60639, tel. 312-252-8220."

Wurlitzer Company

"The Wurlitzer Company makes available a toll-free telephone line, 800-435-2930, to assist any technician or customer in any and all needs pertaining to the Wurlitzer product. The Wurlitzer company maintains a staff of field service managers who can assist should an RFI problem arise." Wurlitzer Co., 403 E. Gurler Rd., DeKalb, IL 60015.

Yamaha International Corporation

The Yamaha organization attempts to cure each RFI problem on an individual basis. Yamaha supplies all necessary technical information at no charge. If interference is caused by design error, Yamaha takes steps at its own expense to remedy the problem. Refer RFI problems to the local dealer. The dealers are kept well informed and current on RFI countermeasures. Inquiries may be directed to Mr. William Perkins, Electronic Service Manager, Electronic Service Dept., P. O. Box 6600, Buena Park, CA 90622, tel. 714-522-9351.

Zenith Radio Corporation

"Zenith gives consideration to handling and providing relief for RFI problems on a case-by-case basis. RFI problems should be referred to Service Division, 11000 W. Seymour Ave., Franklin Park, IL 60131, tel. 312-671-7550. RFI referrals should include model and serial numbers of the affected unit. Customers with a unique, difficult problem may direct a letter to Mr. Richard Wilson, National Service Manager, at the same address."

Other Manufacturers

Ms. Sally Browne, Director of Consumer Affairs, Consumer Electronics Group, Electronic Industries Association, 2001 Eye St., N.W., Washington, DC 20006, tel. 202-457-4900, may be contacted for assistance or recommendations in the handling of RFI problems involving manufacturers not listed here, or for assistance when the product is no longer manufactured.

By Harold R. Richman, W4CIZ

The author is a former FCC Engineer in Charge and is well versed on the subject of RFI. Hal is a member of the ARRL RFI Task Group, and has presented numerous papers on RFI and RFI correction at club meetings, seminars and technical symposia. He was also the recipient of the ARRL Roanoke Division Service Award. His special efforts have been recognized by the ARRL and the FCC, which included Richman's original RFI assistance list in its RFI publication, How to Identify and Resolve Radio-TV Interference Problems. QST

MICROWAVE OVEN RFI

RF heating has been used in industry since 1928 but only in the last few years has it invaded the home. Today, thousands of microwave ovens are sold each year to homemakers who wish to take advantage of quick food preparation provided by this new application of an old idea.

The typical microwave oven operates at about 2450 MHz using a magnetron oscillator working from a rectified, nonfiltered high voltage power supply. The magnetron and oven cavity are well shielded to protect the user, but in most instances the high voltage system is unshielded and the power leads are not effectively filtered for RFI.

If the high voltage diodes in the supply are not properly bypassed, annoying switching transients are generated on each pulse of the rectified ac, and these pulses are radiated and conducted via the powerline to nearby television or radio receivers. The diode transients are heard as a buzzing noise on the radio and show up as a set of wavy lines on the tv screen.

Some oven manufacturers, upon request--and for a price--will supply a suitable power line filter that can be mounted inside the oven cabinet near the input power terminals. A homemade filter, such as the one shown on page 42, Figure 14, can be installed on the oven, provided the oven does not draw more than 600 watts. For greater power capacity, the filter coils should be wound with No. 14 wire.

CORDLESS TELEPHONE INTERFERENCE

Under certain circumstances, some cordless telephones are a source of prolific interference to 80 and 160 meter amateur operators. The phones use a frequency between 1700 and 1800 kHz for one side of a two-way circuit. Some phones are adjusted (or maladjusted) so that the operating frequency is in the 160 meter amateur band, while others generate harmonics that fall in the 80 meter band. The other side of the circuit uses a frequency near 49 MHz. The signals are frequency modulated (fm) and are supposed to be carried along the power line, or are radiated over short distances by a small whip antenna, depending upon the equipment design. Experience has shown that some units are audible over several miles on the so-called 1750 kHz channel which, if improperly adjusted, can cause interference to amateur operators at 1800 kHz.

Some "long-range" units (currently not licensed for operation in the United States) incorporate a power amplifier and a large antenna, and promise "up to 60 mile range". These illegal units can cause severe problems to 6-meter operators, as well as 80 and 160 meter stations. On the other hand, some telephone users report interference from nearby amateur stations whose signals enter the poorly shielded circuits of the cordless phone.

The cordless telephone falls under FCC regulations, as summarized in Part 15, Subpart D of the Rules and Regulations. Paragraph 15.152 states that if an unlicensed low power communication device causes interference to another service, it must cease operation until the interference is eliminated. Any amateur (or other licensed station) receiving interference from a cordless telephone is advised to alert the nearest FCC District Office.

CABLE TELEVISION INTERFERENCE (CATVI)--CONTINUED

Cable television (CATV) consists of a variety of services carried over coaxial cable networks to individual subscribers. The information is displayed on the user's television receiver.

Early CATV provided good reception for fringe area communities. A high-gain antenna was placed at a good receiving site to serve the users via coax cable interconnection between the remote antenna and the residents. Most of these early systems used the standard vhf channels and many of these systems are still in use. Modern CATV installations offer a greater information capacity, having as many as 55 channels. These systems provide a converter to expand the tv receiver capacity, if necessary.

The main path of the CATV system is via trunk cables and trunk amplifiers. The trunk system may extend for many miles and contain dozens of amplifiers. The signal distributed to the end user is extracted from the trunk line and reamplified by a bridge amplifier, often located in the housing of the trunk amplifier. The bridge amplifiers ("bridgers") feed the distribution system, which has branches that supply the subscribers.

The 36 channel CATV system (which encompasses vhf and uhf tv, plus the fm band) occupies the 54 to 300 MHz range while the 55 channel system occupies the 54 to 440 MHz range. These ranges encompass the amateur 144, 220 and 440 MHz bands. The frequency channels used on the CATV system are summarized in Table 1.

CATV "LEAK-OUT" AND "LEAK-IN"

A good CATV system is a closed system. That is, it does not "leak." Signals in the system do not escape and signals outside the system do not enter it. CATVI is caused by a leak in the system. Most system leaks occur in the flexible drop cable from the line to the home. These problems occur because

Table 1 – CATV Channels

Channel Name	Visual Carrier Frequency Standard	HRC	IRC	Channel Name	Visual Carrier Frequency Standard	HRC	IRC
2	55.25	54.0	55.25	L	229.25	228.0	229.25
3	61.25	60.0	61.25	M	235.25	234.0	235.25
4	67.25	66.0	67.25	N	241.25	240.0	241.25
5	77.25	78.0	79.25	O	247.25	246.0	247.25
6	83.25	84.0	85.25	P	253.25	252.0	253.25
A-2	109.25	108.0	109.25	Q	259.25	258.0	259.25
A-1	115.25	114.0	115.25	R	265.25	264.0	265.25
A	121.25	120.0	121.25	S	271.25	270.0	271.25
B	127.25	126.0	127.25	T	277.25	276.0	277.25
C	133.25	132.0	133.25	U	283.25	282.0	283.25
D	139.25	138.0	139.25	V	289.25	288.0	289.25
E	145.25	144.0	145.25	W	295.25	294.0	295.25
F	151.25	150.0	151.25	AA	301.25	300.0	301.25
G	157.25	156.0	157.25	BB	307.25	306.0	307.25
H	163.25	162.0	163.25	CC	313.25	312.0	313.25
I	169.25	168.0	169.25	DD	319.25	318.0	319.25
7	175.25	174.0	175.25	EE	325.25	324.0	325.25
8	181.25	180.0	181.25	.	.	.	.
9	187.25	186.0	189.25	.	.	.	.
10	193.25	192.0	193.25	UU	421.25	420.0	421.25
11	199.25	198.0	199.25	VV	427.25	426.0	427.25
12	205.25	204.0	205.25	WW	433.25	432.0	433.25
13	211.25	210.0	211.25	XX	439.25	438.0	439.25
J	217.25	216.0	217.25	YY	445.25	444.0	445.25
K	223.25	222.0	223.25	ZZ	451.25	450.0	451.25

of poor installation, buffeting of the cable by the wind, or faulty connectors. Water and corrosion take a heavy toll on CATV joints and fittings, and improper ground connections lead to leakage or signal rectification.

Signal leakage can cause havoc to a primary communication service since the CATV channels run through three amateur bands, the aviation bands, and mobile service bands. In particular, channel E falls in the 2 meter band, channels J and K in the 220 MHz amateur band, and channels UU through YY fall in the 432-450 MHz amateur band. Past CATV interference to aviation frequencies has brought FCC rule changes, requiring prior FCC notification before any CATV signal is placed on an aviation frequency. Operation will not be approved if there is an aviation user within 60 nautical miles of any part of the cable system. Unfortunately, radio amateurs (at this writing) enjoy no such protection. In some areas of the country, repeaters on 145.11 MHz have had to cease operations or move frequency because of severe interference from CATV channel E.

CATV "leak-in" is an equally serious problem. Low power amateur operation near 145.11 MHz can ruin channel E reception over a wide area, affecting hundreds of viewers. The amateur, operating within his rights, may bear the brunt of criticism from viewers whose entertainment is disrupted by amateur transmissions.

To make matters worse, unauthorized and illegal connections are sometimes added to the CATV system and legitimate users often supply their own haywire home distribution system. The cable TV operators rightly claim they have no control over what is done with their signals once they reach the customer.

ALLOWABLE CATV SIGNAL RADIATION

Section 76.605 (a) 12 of the FCC Rules sets the limits for allowable system radiation. Up to and including 54 MHz the radiation limit is 15 microvolts-per-meter at a distance of 100 feet from the installation; 54 MHz up to and including 216 MHz, 20 microvolts-per-meter at 10 feet; and over 216 MHz, 15 microvolts-per-meter at 100 feet. Section 76.609 of the Rules explains the method of measurement of these parameters.

Of critical importance to amateurs experiencing CATVI is paragraph (b) of Section 76.613 of the Rules:

(b) The operator of a cable television system that causes harmful interference shall promptly take appropriate measures to eliminate harmful interference.

WHAT TO DO ABOUT CATVI

If an amateur experiences CATVI, he should determine the point of the leak in the system by listening to the interference on a portable receiver. If the leak is found, a letter should be sent to the system operator outlining the problem and stating the location of the leak and the CATV channel causing the problems. Section 76.613 of the FCC Rules concerning the obligations of the operator to clean up the interference should be cited. If the cable company is uncooperative, a second letter should be sent, with copies to the local FCC District Office and the municipal government exercising local control over the CATV system operation.

It is important that the amateur's role in this situation not be that of the prosecutor but that of a diplomat and ambassador for the Amateur Radio Service. The CATV operator will, in many cases, be eager to clean up the problem as he has legal, moral and economic incentives to recognize in satisfying his viewers.

Other groups that can assist in hard-to-solve problems are the National Cable Television Association, 1724 Massachusetts Avenue, N.W., Washington, DC 20036; The Society of Cable Television Engineers, Box 2665, Arlington, VA 22202; and Community Antenna Television Association, 1100-17th Street, N.W., Washington, D.C. 20036. All of these groups have strong ties to both the cable industry and the Federal Communications Commission. Writing the American Radio Relay League, 225 Main Street, Newington, CT 06111, will also prove helpful; the ARRL should be kept fully informed of all cases of CATV interference.

EARTH STATION INTERFERENCE

Microwave dishes are seen more and more often in backyards as enthusiasts enjoy direct pickup of satellite-transmitted television programs with their TVRO ("television receive-only") system. The various satellites, using the 4 GHz band, provide expanded programming for those who go to the trouble of installing their own earth stations. Many new viewers, however, find out that they suffer severe picture degradation on one or more of the 24 available channels due to the reception of unwanted commercial microwave transmissions.

The earth station band (3.7 to 4.2 GHz) is bracketed by, and contains, numerous other microwave services. Most of the interference problems are created by the microwave telephone carriers which mix with the earth station channels. Interference from these signals shows up as colored dashes flashing across the screen, and is referred to as sparklies.

A second source of interference is caused by out-of-band commercial signals that cannot be rejected by the earth station receiver. The unwanted signals mix with the desired signal to produce signal degradation, reducing picture clarity. The microwave common carriers near 2 GHz and 6 GHz are the most common offenders.

ELIMINATING TVRO INTERFERENCE

The interfering signals from the microwave telephone systems are placed between the television signals. A frequency separation of 10 MHz exists between the unwanted and the desired signal at the output of the down converter. (This unit converts the satellite picture to a frequency suitable for the local tv receiver.) The converted picture frequency is 70 MHz (tv channel 4) and two traps in series tuned to 60 MHz and 80 MHz between the converter and the tv receiver will greatly attenuate the offending signals.

Out-of-band interference can often be cured by the addition of a microwave bandpass filter between the low-noise amplifier (LNA) mounted in the feedhorn of the dish and the down converter. This filter passes the TVRO channels and rejects out-of-band signals.

Suitable filters for these purposes are available from Unadilla/Reyco Division of Microwave Filter Corp., East Syracuse, N.Y. 13057, and other manufacturers.

MORE ON TELEPHONE INTERFERENCE

"Trimline" telephone interference can often be cured by wrapping and taping 6-8 turns of the telephone cord around a ferrite rod (Amidon FT-240-43) just before the cord plugs into the jack.

TVI IN APARTMENTS AND CONDOS

If you are operating in an apartment or condominium, these steps will reduce or eliminate TVI:

(1) Always use low power, the lower the better. QRP is best!

(2) Clean up the mess of wires behind the operating table. Keep leads short; use small coax where necessary.

(3) Use an ac-line filter, low pass and high pass filters.

(4) Plug a minimum of accessories into the ac line — use a battery-operated keyer, etc., wherever possible.

(5) Use the radial wires recommended in chapter 9.

(6) A random length wire antenna, fed through a tuner, produces more TVI than other types. If you use an indoor antenna, changing its position somewhat may reduce or cure TVI.

(7) One amateur found that unplugging his TV set, even though it was "Off," cured a neighbor's TVI.

GARAGE DOOR INTERFERENCE

Garage door openers can create TVI, and also be activated by a strong rf signal. The leads to the auxiliary control switch should be filtered at the receiver input by series-connected 20 uH chokes. Bypass each lead to the receiver chassis with a .01 uF disc capacitor.

VCR INTERFERENCE

This is a complex problem. RFI is worst on 80 meters. Never run more than 100 w on cw or 250 w on ssb — the lower the better. Antenna lead to VCR, and lead from VCR to TV set, should be wrapped around ferrite rods. Wrap ac-line leads to VCR and TV set around ferrite cores. Use high pass filters before VCR and between VCR and TV set. Cheap VCRs have RFI even after these steps because they omit shielding to cut costs.

About the Author

William R. Nelson, WA6FQG, joined the Southern California Edison Company as a groundsman in 1947 and became a lineman in 1949. Following five years as a lineman, he was advanced to estimator. His work included distribution design, power facilities and load management. In 1964 he was appointed Amateur Radio Representative and RFI Investigator for the company and held that position until his retirement in 1980.

During his 16 years as an Investigator, Bill Nelson tracked down countless cases of RFI and lectured at amateur radio clubs and conventions on the causes and cures of interference from power facilities and other sources. He was instrumental in bringing about changes in construction practices to correct and eliminate power line RFI. Many of these changes are utilized today in the construction practices of other electric utilities.

Bill Nelson is past-Chairman of the Los Angeles Council of Radio Clubs' TVI Committee. He is now a consultant to power utilities onRFI problems, including investigation and training. He is uniquely qualified to author this comprehensive reference work on all types of interference.

* * * *

INDEX

OTHER BOOKS FOR RADIO AMATEURS, CB OPERATORS, SHORTWAVE LISTENERS, STUDENTS, & EXPERIMENTERS

ALL ABOUT CUBICAL QUAD ANTENNAS, by William I. Orr W6SAI and Stuart D. Cowan W2LX; 112 pages, 75 illustrations.

This well-known classic has been updated to include: new Quad designs; new dimension charts for every type of Quad from 6 to 80 meters; additional gain figures; an analysis of Quad vs. Yagi; Mini and Monster Quad designs; Delta, Swiss, and Birdcage Quads; and an improved Tri-Gamma match to feed a triband Quad efficiently with one transmission line. Also covered are feed systems and tuning procedures for maximum gain and minimum SWR. Much of this data has never before been published.

BEAM ANTENNA HANDBOOK, by William I. Orr W6SAI and Stuart D. Cowan W2LX; 271 pages, 205 illustrations.

This popular new edition gives you: correct dimensions for 6, 10, 15, 20, and 40 meter beams; data on triband and compact beams; the truth about beam height; SWR curves for popular beams from 6 to 40 meters; and comparisons of T-match, Gamma match, and direct feed. Describes tests to confirm if your beam is working properly, tells how to save money by building your own beam and balun, and discusses test instruments and how to use them. A "must" for the serious DX'er!

THE RADIO AMATEUR ANTENNA HANDBOOK, by William I. Orr W6SAI and Stuart D. Cowan W2LX; 191 pages, 147 illustrations.

This clearly written, easy to understand handbook contains a wealth of information about amateur antennas, from beams to baluns, tuners, and towers. The exclusive "Truth Table" gives you the actual dB gain of 10 popular antenna types. Describes how to build multiband vertical and horizontal antennas, Quads, Delta Quads, Mini-Quads, a Monster Quad, DX "slopers", triband beams, and VHF Quagi and log periodic Yagi beams. Dimensions are given for all antennas in English and Metric units. Tells how antenna height and location affect results and describes efficient antennas for areas with poor ground conductivity. Covers radials, coaxial cable loss, "bargain" coax, baluns, SWR meters, wind loading, tower hazards, and the advantages and disadvantages of crank-up tilt-over towers.

SIMPLE, LOW-COST WIRE ANTENNAS FOR RADIO AMATEURS, by William I. Orr W6SAI and Stuart D. Cowan, W2LX; 192 pages, 100 illustrations.

Now–another great handbook joins the famous Cubical Quad and Beam Antenna Handbooks. Provides complete instructions for building tested wire antennas from 2 through 160 meters–horizontal, vertical, multiband traps, and beam antennas. Describes a 3-band Novice dipole with only one feedline; the "folded Marconi" antenna for 40, 80, or 160 meters; "invisible" antennas for difficult locations–hidden, disguised, and disappearing antennas (the Dick Tracy Special, the CIA Special). Covers antenna tuners and baluns, and gives clear explanations of radiation resistance, impedance, radials, ground systems, and lightning protection. This is a truly practical handbook.

THE TRUTH ABOUT CB ANTENNAS, by William I. Orr W6SAI and Stuart D. Cowan W2LX; 240 pages, 145 illustrations.

Contains everything the CB'er needs to know to buy or build, install, and adjust efficient CB antennas for strong, reliable signals. A unique "Truth Table" shows the dB gain from 10 of the most popular CB antennas. The antenna is the key to clear, reliable communication but most CB antennas do not work near peak efficiency. Now, for the first time, this handbook gives clear informative instructions on antenna adjustment, exposes false claims about inferior antennas, and helps you make your antenna work. With exclusive and complete coverage of the "Monster Quad" beam, the "King" of CB antennas.